Winterhaven Townsite Company

Imperial County, California

xxx

...OMPANY

ARIZONA
ARIZONA

COCHISE DEVELOPMENT COMPANY

WINTERHAVEN EGYPTIAN COTTON RANCH

A. E. DEYO, Manager

The Winter Garden Section of Southern California

WINTERHAVEN TOWNSITE COMPANY

INCORPORATED UNDER THE LAWS OF CALIFORNIA

Denn Arizona Copper Company

Bisbee, Arizona

COMBINATION GOLD MINING COMPANY

761-764 BOARD OF TRADE BUILDING

...ANSAS CITY, MISSOURI.

GREAT WEST AND INDIAN SERIES, VOLUME FIFTY-SIX

LEMUEL C. SHATTUCK

"A LITTLE MINING, A LITTLE BANKING, AND A LITTLE BEER"

To Ann;
Thank you for sending me
on my way to a finished book
Isabel Shattuck Fathauer

LEMUEL C. SHATTUCK

"A LITTLE MINING, A LITTLE BANKING, AND A LITTLE BEER"

By

ISABEL SHATTUCK FATHAUER

With additional research
and editorial assistance
by

Lynn R. Bailey

WESTERNLORE PRESS . . . 1991 . . . TUCSON, ARIZONA

Library of Congress Catalog Number 91–75242
ISBN 0-87026–079–0

PRINTED IN THE UNITED STATES OF AMERICA BY WESTERNLORE PRESS

To my husband, Walter Fathauer,
who supported me in my many
interests for fifty-nine years.

TABLE OF CONTENTS

Table of Contents Continued

INTRODUCTION

WHEN asked by a judge to state his occupation, Lemuel Coover Shattuck replied in a matter of fact manner: "A little mining, a little banking, and a little beer." While some laughed at the tongue-in-check, self-effacing answer, all in the Tombstone courtroom knew this mustached Pennsylvanian who sat bear-like in the witness stand. He was a booster of Cochise County, a founding father of Bisbee, and one of the West's most knowledgeable mining men. There was nothing in Shattuck's answer to indicate he was one of Arizona's wealthiest and most influential men. Yet three decades of hard work is reflected in his succinct remark.

Lem, as Shattuck was affectionately called, was seventeen-years-old when he arrived in Arizona Territory in 1883 to join the Erie Cattle Company, a Sulphur Springs Valley ranching enterprise owned by his half-brothers. Four years of chasing cattle and avoiding Apaches convinced the youth there was an easier way to earn a living.

Bidding farewell to his half-brothers, Lemuel drifted into Bisbee, a budding copper camp in the Mule Mountains, and took a job at the Copper Prince smelter. His employment was of short duration; closure of the Prince Mine forced Lemuel to hit the road in search for his fortune. For two years he drifted around the West working as a miner and a teamster, and occasionally prospecting. In 1888 he returned to Bisbee — penniless, a bedroll on his back.

Shattuck may have been down on his luck, but he was in the right place at the right time. The Electric Age had begun and the demand for copper was insatiable; Bisbee was about to boom. With instincts of a gambler, Lemuel sensed and grasped the opportunity at hand. Laboring at night in the dark drifts of the Czar Mine, he made adobe bricks on a plot of ground in upper Brewery Gulch during the day. As mining expanded so did the need for shelter, and the brickyard soon became the Shattuck Lumber Company.

Shattuck systematically built in other ways. In partnership with others, he founded the camp's first municipal water works, erected a skating rink, an opera house, and a number of substantial commercial buildings. Armed with an Anheuser-Busch Beer franchise, Lemuel distributed keg beer throughout Cochise County, and as far south as Cananea, Mexico. His St. Louis Beer Hall, later renamed the Shattuck Saloon, is remembered to this day as one of Bisbee's finest masculine retreats.

As a businessman, Lemuel Shattuck has been described as a plodder. A juggler is more appropriate. As he dealt in booze and lumber, he aggressively sought mining property: speculating in Mexican mineral rights, and staking claims throughout the Warren Mining District, as the copper producing area around Bisbee was called. He bought some claims, others passed into his possession with the turn of cards.

Mine development required capitalization. But money was in short supply in a region far removed from centers of Eastern finance. Shattuck sought to remedy that deficiency by founding a bank. Conceived to aid Bisbee's commercial community and emerging mining companies, his Miners and Merchants Bank was an immediate success, due to an aggressive, yet open-minded approach to the region's banking needs. The door to Lemuel's office was never closed to those in need. He dealt in good faith, often with a handshake, and expected others to do the same, to the benefit of the little man as well as the powerful.

The Miners and Merchants Bank gave Shattuck a foundation on

which to build, both locally and throughout the West. The Bank financed Bisbee improvements, southeastern Arizona ranchers and businessmen, Yuma Valley cotton farmers, and Salt River Valley diarymen. Shattuck underwrote much of the development of Cochise County. As at Bisbee, his money went into the building of Douglas. And Lemuel planned and financed Winterhaven, California, opposite Yuma. It was Shattuck who backed the first commercial crop of Egyptian extra-long staple cotton produced in the Colorado River Valley.

The Bank established Shattuck's creditability with Eastern financiers, allowing him to turn claims into paying mines. He sold mineral rights to Midwestern speculators. Some of his claims were purchased by development companies that eventually became the Calumet and Arizona Mining Company. At one time Shattuck owned a portion of the ground exploited by the Junction Mine, one of Bisbee's great copper producers. His greatest faith, however, lay in a group of claims in a canyon high on the mountain overlooking Bisbee. With financial backing from Minnesota and Michigan mining men, Shattuck developed his "Richest Little Mine." Producing the richest ore, at the lowest production cost of any copper mine in the world, the Shattuck-Arizona Mining Company paid over $8 million in dividends in two decades. Its money underwrote development of another Bisbee copper producer, the Denn-Arizona Mining Company. Shattuck also dabbled in gold, procuring the financial backing for the Lucky Tiger-Combination Gold Mining Company in Sonora, one of the world's great gold producers.

A pioneer builder without equal, Lemuel Shattuck never flaunted success, or abused power. The term "Robber Baron," often used in connection with Western mining men, cannot be applied to this man. Although he owned the controlling share of Bisbee's third largest mining company, one of Arizona's soundest financial institutions, and had other business interests worth millions of dollars, Lemuel Shattuck lived modestly in the town where he made his fortunes. He never lost his rugged individualism and integrity. He could sit in executive boardrooms, surrounded by financiers from New York, Boston, and Duluth, and plan gigantic developments costing millions of dollars, yet he spoke the language of the sourdough and humble prospector. Gruff, straight talking, frank and outspoken, Lemuel's heart was near the surface.

Builders are endowed with a sense of destiny, and usually let the world know it. Shattuck had a destiny and knew it, but did not advertise it. He quietly built—not only for himself and stockholders of his companies—but for his family and his community. Lemuel's modesty and reticence were endearing to people close to him, yet maddening from the standpoint of history. A raconteur whose stories were never recorded, he was interviewed only twice in his lifetime. Shattuck shunned public appearances. His style was to work in the background.

Passage of Cochise County stalwarts focused attention on Lemuel Shattuck. In the 1930s journalists and authors frequently called on him, requesting interviews. Being a private man with a basic humility, Lemuel declined with one exception. He granted an interview to Walter Zipf in 1934. Published in the Bisbee *Daily Review* on May 20 under the title "Lem Shattuck: The Copper Country Commoner," this lengthy biographical sketch formed the basis for all subsequent treatments of Shattuck. All who knew Shattuck recognized that Zipf barely scratched the surface. Others volunteered to write his history. Lemuel always declined. When approached by his daughter, he dropped his aloofness.

With a full-scale biography in mind, Isabel Shattuck Fathauer hung on to her father's personal papers, and began the search for additional material that would cast light on his life. The quest, spanning many years, has been aided by numerous individuals and institutions. Both the Arizona Historical Society and the University of Arizona Library gave access to their archival holdings, runs of early-day newspapers, mining periodicals, and journals. Tom Vaughan of the Bisbee Mining and Historical Museum furnished photographic materials and advised on aspects of Shattuck's life. He, along with Harwood P. Hinton of the University of Arizona, and C. L. Sonnichsen of the Arizona Historical Society, aided Dan Shattuck in gathering data on the Erie Cattle Company. Alice

Metz, faithful volunteer at the Shattuck Library, graciously provided bibliographic notes taken from the Bisbee *Daily Review*. Joseph Muheim, Jr., recalled tales of his father's association with Shattuck. Various departments of Cochise County also furnished records and statistics. The Pennsylvania State Archives, and the Erie Historical Society likewise aided in the search for geneological material. A number of individuals, however, turned a dream into reality.

John Gilchriese, former Field Historian of the University of Arizona, urged the undertaking of a Shattuck biography. Lynn R. Bailey of Westernlore Press guided the editorial preparation of the manuscript and enlarged the book with additional research. Dan Shattuck, of Ashland, Kansas, graciously provided material on Enoch and Jonas Shattuck, and the Erie Cattle Company. John Bret Hart offered guidance in the early stages of the work, and Ann Trautman computerized initial drafts. Last, but certainly not least, geologist Weldon Humphrey, Jr., urged the inclusion of geological and mining data in any biography of Lemuel Shattuck.

Isabel Shattuck Fathauer
Tucson, Arizona
June 10, 1991

Lemuel Coover Shattuck was born January 5, 1866, in Erie, Pennsylvania. At age seventeen, shortly after this photograph was taken, he joined his half-brothers' ranching enterprise in the Sulphur Springs Valley of southeastern Arizona. (Photo courtesy Dan Shattuck)

ONE

THE ERIE CATTLE COMPANY

IN the fall of 1883 seventeen-year-old Lemuel Coover Shattuck stepped off the Southern Pacific train at Benson, Arizona Territory. The trip from Erie, Pennsylvania, was long, nearly ten days: four days and three rail connections to reach New Orleans, and six days and two train changes to cross Texas, through cattle country he heard so much about. Tweed clothes and a tie announced Shattuck as a greenhorn Easterner. Large for his age, just under six feet, big-boned, brown-haired, and blue-eyed, he had a degree of stoicism about him that could be mistaken for western characteristics. Reticence and matter-of-factness were traits common to both Arizonans and Pennsylvanians. While saddened about leaving home, this was the moment he had longed for—the moment in every young man's life when he takes his first step on the path of dreams. And Lemuel's dream had always been to work with his half-brothers in their ranching business.

Like his father and half-brothers, Lemuel was a farmer's son. In 1814 his grandfather Spencer had left Torrington, Connecticut, and moved westward beyond the Appalachian Mountains, seeking land under provisions of land grants and warrants that gave enterprising frontiersmen opportunities to acquire land for as little as $1.50 an acre. In northwestern Pennsylvania on the shore of Lake Erie, between Walnut and Elk creeks, he staked his claim; cleared and planted the land to corn, wheat, and potatoes; and raised a large family. After Spencer's death in 1852, his son Henry carried on with help of his sons William, Enoch Austin (nicknamed "Aus"), and Jonas (called "Stub" because he was the tall one), and a daughter, Irene.[1] The Shattuck land was fertile and the family prospered. Henry built a grist mill on Elk Creek, expanded into buying and selling grain, and acquired a sizable dairy herd.

There were sad times too. Henry's first wife, Emily (whom he had married in 1817), passed away within a month of Spencer's death; and Henry spent a lonely eight years before remarrying. In 1860 he married Phoebe Ann Coover, a second-generation Pennsylvanian of Palatine-German descent, who bore him three children: Lemuel in 1866, Eldridge in 1867, and John in 1872.[2] The latter two did not live into the 20th Century, and knowledge of Lemuel's early childhood is fragmentary.

Like all the Shattuck children, Lemuel received an education in care of dairy cattle, horses, and farm work. Planting and harvesting duties, however, were somewhat cut short by incorporation of a portion of the family farmland, including the original brick farmhouse built by Spencer, into the Erie municipality. Undaunted, Henry built another home in town — on the corner of 24th and German Streets — the exact replica of the old farmhouse, only more elaborate within. Behind the structure he planted a garden and an orchard. Into this house Henry settled Phoebe and their children.

Northwestern Pennsylvania prospered with opening of the Erie Extension Canal and the coming of the Sunbury and Erie Railroad. Iron and coal deposits were opened, and by 1857 Erie had become a smelting and shipping center with a population exceeding 38,000.[3] The Civil War bolstered the economy further. Progressive and farsighted, Henry took advantage of the good times by concentrating on the purchase and sale of grain and livestock. By end of the Civil War supply of beef and horses could not keep pace with demand, and Henry was forced to canvas for livestock throughout western Pennsylvania and into Ohio. His trade in cattle and horses boomed, and by the 1870s Henry and his two sons, Aus and Stub, were looking for markets far afield, in Kansas, and beyond.

The Kansas Pacific Railroad had extended its rails to the eastern edge of the Cherokee Strip, spurring a lucrative trade in buffalo hides, as well as cattle driven north from Texas over the Chisholm Trail. In 1878 the two sons, financed by their father, and perhaps other Erie men, ventured into Kansas and the Cherokee Strip to buy cattle and horses. At first they brokered livestock, purchasing longhorns from Texas drovers at Ellsworth and other Kansas railheads, and then loading them — sixteen animals to a cattle car — for shipment by the Kansas & Pacific Railroad to eastern markets. Profit lay in fattening cattle prior to shipping them East, and the Shattucks filed a claim to 160 acres along the Ninnescah River in Kingman County, in south-central Kansas. By 1880 they had a modest herd of about 300 animals, which they sold that fall. In partnership with Thomas Roland, and Edward and W. S. Parker, the Shattucks headed for Fort Worth, Texas, where they purchased 750 two- and three-year old steers, at seven dollars a head, which they drove up the Chisholm Trail to Indian Territory. Fattening the cattle on the rolling, well-watered terrain about Pond Creek, the Shattucks sold the herd for fourteen to fifteen dollars a head during summer of 1882. Flushed with money they again headed for Texas, purchased 1,400 steers, which were again driven to Indian Territory and subsequently sold. By buying, fattening, and selling cattle, the Shattucks increased their capital ninefold — in the short span of five years they mastered the range cattle business.[4]

Meanwhile, Henry had his hands full in Erie. The move into town proved difficult for his youngest son. Lemuel missed the farm. The lush green acreage, rolling up from the stream to wooded highlands beyond, had been his playground. Swirling waters of the millrace held a fascination broken only after a close call when he fell in. He was saved by his dog, who jumped in after the child, sunk his teeth in the little fellow's red smock and pulled him to safety. After that narrow escape, Lemuel was more careful about approaching swift water.

Some aspects of town life appealed to the youth. Judging from his adult correspondence, he excelled in grade school, receiving a

Henry Shattuck, father of Lemuel, Enoch, and Jonas, was a prominent Erie farmer, dairyman, and merchant. (Photo courtesy Dan Shattuck)

thorough foundation in the "three Rs," especially mathematics. When not in school or working for his father, the wharves and docks along the lake front held his attention. He watched the passage and docking of vessels carrying coal and iron ore; and he fished for whitefish, the taste for which he kept all his life. He taught himself to swim and grasped the art of sailing, finding joy in skimming over the water powered only by the wind. Above all, he acquired his father's Universalist ethics of hard work and fair dealing.

Town life increased the gulf — already wide due to a fifteen year age difference — between Lemuel and his half brothers and sister. The family's livestock enterprises took Jonas and Enoch away from Erie much of the time during the late 1870s. Their comings and goings were a mystery to the young boy, and Lemuel anxiously awaited their visits. Aus and Stub were cowboys, bronzed by long days on the range, quiet but observant. They walked as if they had spent a lifetime in the saddle, and their stories of cattle drives and cowboys and cattle towns thrilled the young boy. Lemuel yearned for such a life.

The time came when Lemuel would realize that dream. By 1880 cattlemen of the Cherokee Strip were an insecure lot — little more than tenants on Indian land, enjoying grazing privileges only so long as they paid taxes to the tribe. And those taxes were on the increase. Cattlemen of the Strip also faced the threat of encroachment by outsiders, who were gathering along the Kansas line in anticipation of government action to open Indian Territory to settlement. The range cattle business in Kansas was being constricted by fences and regulations prohibiting importation of cattle infected with Texas fever, a lethal splenic disease induced by ticks, which could be passed to other livestock.[5] As their business in the Cherokee Strip faltered, the Shattucks literally looked for greener pasture.

Montana and Wyoming appeared promising, but winters were long and bitterly cold. Texas was out of the question. It, too, was being fenced and legislation there was detrimental to new ranchers. They heard talk of land beyond the Rio Grande, in New

A product of a quiet Pennsylvania rural upbringing, Enoch Austin Shattuck was a man of quiet demeanor, and of integrity. Although he never fired a shot in wrath or self-defense, "Aus" served as undersheriff to Cochise County Sheriff John Slaughter, between 1886 and 1890. The two men maintained a close business and personal relationship for more than ten years. Enoch was in his early forties when this photograph was taken. (Photo courtesy Dan Shattuck)

Mexico and Arizona territories, where the country was arid but the climate conducive to cattle raising. Above all else, the area had great economic potential. Mineral wealth abounded between the Rocky Mountains and the Pacific Coast. Mining camps mushroomed, and military posts and Indian reservations were being established, all outlets for enterprising ranchers. And railroads were spanning the continent, linking resources to consumers. Just where to locate was the question the Shattucks had to resolve.

A number of cattlemen had chosen Arizona and New Mexico territories. In the late 1860s Charles Goodnight, John Simpson Chisum, and Oliver Loving drove Texas cattle in New Mexico. They were followed by Charley and John Slaughter of Friotown, Texas, who settled in the Guadalupe Mountains. John Slaughter, however, did not stay in New Mexico. He had been in southeastern Arizona as early as 1877 and liked what he saw. A year later he pushed across the Rio Grande and drove his herd on to the Sulphur Springs Valley in southeastern Arizona. In 1884 John Slaughter purchased from G. Andrade 65,000 acres comprising the San Bernardino land grant located at the southern end of the Sulphur Springs Valley — the only all-grass valley in Arizona.[6]

The Sulphur Springs Valley is a series of basins about twenty miles wide and more than a hundred miles long, extending into Mexico. Its waters drain in opposite directions, part flowing south into the Yaqui River, and part running north through Arivaipa Canyon into the San Pedro River. There gramma grass grew to a man's knees; and with an altitude of from 3,900 to 4,400 feet, it had no extremes of heat and cold. Most important, it was said to have an inexhaustible supply of artesian water rising to within a few feet of the surface. The valley was ideal for cattle; in fact, it teemed with wild cattle, the remnants of Spanish and Mexican ranchos put out of business by raiding Apaches.[7]

Whether their choice was prompted by conversations with Slaughter or other cattlemen, through careful perusal of government documents, on-site inspection, or even all three, no one knows. What is known is that the Shattucks settled for the Sulphur Springs Valley; and on August 15, 1883, Enoch together with as-

Because Jonas Henry Shattuck was more than six feet tall he was called "Stub." Unlike older brother Enoch, Jonas retired from ranching to run a lumber business in his home town of Erie, Pennsylvania.

(Photo courtesy Dan Shattuck)

sociates from Erie, Pennsylvania — Benjamin F. Brown and Hugh H. Whitney — filed papers of incorporation of the Erie Cattle Company with the Cochise County clerk at Tombstone. Capitalization of their enterprise was $80,000 and the place of business was listed as the town of Willcox, a shipping point on the Southern Pacific Railroad. Henry and Jonas Shattuck also had an interest in the Erie Cattle Company, as did Whitney's son Wallace. There were also a number of minor partners: Milton R. Chambers, Albert G. Smith, and James B. McNair, all from northwestern Pennsylvania.[8]

A never-failing spring of good water in sufficient volume to supply a ranch in range country is a rare occurrence. Such a spring is usually found at long intervals, and is always a desirable and valuable property. It makes an oasis in the desert, and if properly utilized its owner can enjoy a comfortable subsistence for himself and his herds. The isolated rancher who is well located is independent. He is in no danger of being crowded by neighbors nor having his range overstocked with stray cattle. His water rights give him undisputed control of adjacent range, even though he does not own all the land; that is the unwritten law of the range, respected by all cattlemen.

Consequently, the Shattucks and their associates, in accordance with Federal land laws, preempted a number of artesian springs and other water sources.[9] The actual number of acres constituting their initial claim was not large: at most 2,300 to 2,500 acres.[10] Their holdings ran from within the Chiricahua Mountains in a broad arc following Whitewater Draw, south and west to the Mexican border where Douglas now stands. Mud Springs, Double Adobe, and Silver Creek were headquarter locations. The holding of water rights established the Erie's open range, covering thirty to forty miles of the Sulphur Springs Valley, almost the entire valley. Tombstone was their mailing address and the Tumbling T their brand.

After securing the land, these Pennsylvanians hired help, built temporary shelters, and went to work opening their water sources. In the southern basins of the Sulphur Springs Valley much of the water table is within thirty-five feet of the surface. They increased the flow of springs and seeps by simply running a fence post auger into the aquifers and inserting a four-inch pipe. Resultant flow was channeled into reservoirs for stock consumption. Additional water was discovered near thickets of mesquite and cottonwood. And water was always close to the surface where low mounds of black organic material existed.

Once permanent water was secured, Erie management got on with the business of ranching. They searched the range and rounded up every cow, calf, and heifer they could find, those without brands were marked. Those with brands returned to their rightful owners at roundup time. Additional cattle were purchased from other southern Arizona ranches, and from the Snake Ranch at the head of the San Pedro River, owned by the wealthy Camou family, who had banking interests in Guaymas and Hermosillo, Mexico.

When "Aus" extended the invitation to Lemuel to join the Erie outfit, the youth was ecstatic. His ancestors had farmed in America since 1640, but herding and marketing cattle sounded more romantic and glamorous than irrigating rooting plants. In the miles and miles of grasslands of the Sulphur Springs and San Simon valleys Lemuel would serve his cowboy apprenticeship.

At Benson Lemuel stepped into another world. In ten days he had passed from the settled, civilized world of the East to the frontier of the Far West. After he viewed the open, seemingly limitless range of Texas, New Mexico, and the Sulphur Springs Valley, the small fenced fields of northwest Pennsylvania and his father's pasture lots of a few acres dwindled to insignificance. The family's farm could never be restored to its former greatness.

Gone were the luxuries of the Erie townhouse. Ranch headquarters at Double Adobe, six miles southwest of Bisbee, was only a two-room adobe, its three-foot-thick walls slotted with gun ports. The structure, standing since the 1870s, had been refurbished by his half-brothers. It had no unnecessary improvements. Every dollar spent in repairs had been put where it would do the most good. Furnishings were of the plainest kind — handhewn wooden tables, chairs, and beds; cast iron frying pans and dutch ovens, and Papago earthenware. While the larder was adequate, the cuisine included no dainties. It was a "stag camp," composed entirely of men.

Aus and Stub showed no favoritism. Lemuel had joined a man's world and he was treated accordingly. Well almost. At first he did not bed with seasoned cowhands; he was quartered in a dugout, perhaps the temporary shelter used by his brothers when they took up the land. Like everybody on the ranch he shared in the work: from preparing meals and keeping house to herding.

Donning a broad sombrero, blue flannel shirt, fringed chaperejos and jingling spurs, he spent long hours riding the range, marking condition of cattle, retrieving strayed stock, keeping springs and waterholes clean, and branding calves. In fact, there was always branding, for during Erie's formative months, its personnel marked an animal whenever and wherever it was found, not just during spring and fall roundups.

Lemuel worked alongside veteran cowboys who instructed him to keep close watch on newly purchased cattle, driven in from other ranges. Cattle turned loose upon new range are inclined to scatter, and until they became accustomed to the change in bedding ground they were "close herded." Once located they were not liable to stray far.

At first cowboying suited Lemuel and he adapted to ranch life, although he did suffer some disillusionment. In Pennsylvania he thought he knew horses. That was not so. Western horses were smaller and quicker than large Erie farm horses. A cowpony had amazing intelligence, and when directed to a single animal in a herd of cattle, it fixed upon the animal and seldom failed to bring it out. If the reins were dropped to the ground, the western horse stood patiently. If its rider lassoed an animal to be tied, branded or medicated, the horse positioned itself so that the rope tied to the saddle horn remained taut. Such action caused Lemuel to ponder who was smarter, the horse or the cowboy who so perfectly trained it.

Although he knew the habits of the family's dairy cows at Erie, Arizona cattle were wild animals. Livestock on the Sulphur

Springs Valley ranges were worked by men on horseback, and were not frightened at sight of horse and rider. But let a stranger approach on foot, and every head was raised in surprise and alarm. In the ensuing wild stampede the stranger is lucky if he is not run down and trampled underfoot. Rounding up cattle and searching for calves could be both difficult and dangerous. Lemuel learned to look in the areas of maximum cover where cows would hide their calves. Even when closely approached lone calves would not move. To Lemuel they seemed to convey a "not at home attitude." A calf in the wild believed it was invisible, and could be scarcely aroused or frightened away. Its behavior was different from the shy attitude of calves in a herd. The herd instinct was apparent to Lemuel when he observed a group of calves left in the charge of a young cow or heifer that seemed to understand her responsibility to guard her charges carefully. There was, however, a monotonous rhythm to cattle. Paths converged from every water hole over which cattle traveled. With great regularity they moved back and forth from water to grass, going to water in the morning and back to feeding grounds at night.[11]

While Lemuel observed and learned, he in turn was being watched by seasoned cowhands. This greenhorn was somehow related to owners of the outfit, went the rumor. Just how, no cowboy dared ask. Veteran cowhands kept their mouths closed, did their work, and taught the greenhorn to put the well-being of the ranch, the owners and the cattle he tended above his own welfare. Not at all an unreasonable doctrine, and Lemuel complied, worked hard, learned fast. His youthful stamina and strength amazed the cowhands, and they affectionately gave him the inevitable nickname, the "Kid."

Lemuel found the cowhands a diverse lot: experienced cowboys from Texas and New Mexico, younger sons from England with no inheritance, Civil War veterans, midwesterners and neophyte youths like himself. There were also some hard cases. Andy Darnell was a man "not to be trifled with." On his time off he headed for the tenderloin of the nearest town, invariably drank too much and ended up in jail. Eventually he was shot to death at the Leslie

Ranch by Frank Johnson, another Erie cowboy.[12] And there was Will Carver, alias G. W. Franks, a leading member of Black Jack Ketchum's gang. "A nice enough fellow, but melancholy." After leaving the Erie he joined Sam Ketchum's gang, and eventually wound up riding with Butch Cassidy's Hole-in-the-Wall gang. He was killed by Sheriff Briant, at Sonora, Texas, on April 2, 1901.

Lemuel made friends without regard to background, and many of these friends lasted a lifetime. Most of these men displayed a dry sense of humor and often related amusing tales about past experiences. They believed in and practiced the same moral code he had learned in his childhood — decency, fair play, hard work and trust in a man's spoken word unless he proved dishonest and undependable. Every cowboy was accepted at face value, according to his will and ability to do the job. Like farmers, cowhands had a wealth of knowledge about nature, for they had familiarized themselves with terrain, vegetation, and the animals indigenous to the area.

Such knowledge, with its strange sounding words, Lemuel absorbed like a sponge. He learned to recognize loco weed, the barium content of which could send a cow crazy and even kill it. After learning some Spanish, he realized that "Malapai," a name for a rough volcanic area just west of Skeleton Canyon, was a corruption of the term *mal pais* or bad land. What he knew as a creek was called "crick" by Arizonans. He acquired knowledge of the nutritive value of native herbs, grasses and other vegetation, and was amazed to see cattle feeding on mesquite beans. Wild pepper and mustard, as well as certain ground covers that grew profusely during the summer, could be tasted in a cow's milk if she fed on them.

Arizona's animal world was unlike anything he had seen: the family life of quail seemed exemplary: parents ever watchful and protective, mother in front trailed by her chicks, father bringing up the rear. One night he came upon a band of javelinas bedded down for the night. The males had formed a ring around the females and their young. As he urged his reluctant horse nearer for a closer look, the males charged fiercely, causing him to withdraw. And he developed a tolerance for reptiles, learning by experience that un-

Above: The ranch of Wallace W. Whitney at Mud Springs, sometime in the late 1880s or early 1890s. Below: Erie Cattle Company roundup near Box Canyon. Date of photo is unknown, probably in the 1880s.

(Both photos courtesy Dan Shattuck)

less molested or surprised, a snake prefers to slither away to safety. He knew a striking snake could not reach its target if a person was at least the creature's length and a half away. He discovered that reptiles were beneficial, killing packrats and other small rodents. Yet, if necessary, he could kill a rattlesnake with a single shot from his pistol or the throw of a few stones. Like all cowboys, he learned the pitfalls of the open range. Prairie dogs, charming little creatures that popped up to sit and watch everything going on, ate rangeland browse. Their mounded villages were hazards every cowboy knew well. A step into a burrow could break a horse's leg and injure its rider.

Its livestock initially obtained from Mexico and other Arizona ranches, the Erie Cattle Company grew rapidly. According to an entry in the July 17, 1885, issue of the *Daily Tombstone*, the Erie was running about 4,300 head. Three other companies in Cochise County were larger: White and Vickers were listed as having 6,100 head, the San Simon Cattle Company ran 4,500 head, and Tevis, Perrin Land and Cattle had 6,000 animals. There were smaller outfits: Lyle and Sanderson (2,000 head), Hall and Aston (2,000 head), and the Persley Brothers (1,195 head). These and other companies competed for markets in Arizona Territory. Burgeoning towns of Tucson, Tombstone, and Bisbee took their share of cattle. The United States government, however, was the biggest buyer of beef. The military at Forts Huachuca, Bowie, Grant, and Crittenden let large contracts for cattle. In 1881 the Bureau of Indian Affairs alone purchased more than three million pounds of beef for its San Carlos agency.

Government agencies constantly published requests for livestock bids in Tombstone and Tucson newspapers, contracts being contingent upon price and ability to post a bond of several thousand dollars, a sum only the largest outfits could afford. Such contracts were not always profitable, due to niggling buyers and the pitfalls of trail and weather. On January 10, 1885, the *Southwestern Stockman* reported that "out of eighty head of cattle driven through Willcox last week from the Erie Cattle Company's range, and destined for the San Carlos Agency, only twenty reached their destination. The continued storms then raging through the country made it impossible to herd them, and they kept straggling off until only the number mentioned remained." The same weather conditions also prevented Jonas Shattuck from disposing of fifty head of work steers. Heavy snowfall had isolated Globe and other mining locales, and he was unable to contact teamsters in need of animals. According to the *Southwestern Stockman*, he returned to his range without disposing of a single steer. There were other hazards, too, among them the menace of Indians.

Like other ranches of the Sulphur Springs Valley, the Erie Cattle Company literally sat astride a trail—a plunder trail. For generations this and other north-south trending valleys served Western Apache groups as conduits to the mines and ranchos of northern Mexico. The western Coyotero Apaches had two roads that entered Sonora and bore along the Pacific slopes of the Sierra Madre Occidental. These trails, collectively labeled the "Great Stealing Road"—and it was just that, for it was many feet wide in places— ran down Arivaipa Creek and turned southward across the San Pedro Valley, passing what is now Bisbee, to enter Sonora northwest of Fronteras. From that region, thick with ranchos, the trail branched: southwest to the ranchos and mines around Magdalena and southeast to the Opata villages of Bacoachi and Nacozori, or straight ahead to Hermosillo and Arizpe.

A third trail, used by eastern Coyotero Apaches, crossed the Gila farther upstream, led along San Simon Creek and dropped through the San Bernardino land grant to penetrate the Sierra Madres, where it then branched toward both Chihuahua and Sonora. Although establishment of military posts and reservations in southeastern Arizona curtailed Apache raiding following the Civil War, Indians still slipped away to follow their raiding proclivities; if not alert, ranchers whose holdings were situated in the path of southward-moving Indians could pay a high price in stolen livestock or possibly loss of life.

Lemuel Shattuck had been in Arizona Territory less than a year when he ran afoul of Apaches. While riding the range, his horse had thrown a shoe and he repaired to the O'Keefe Ranch at the entrance of Rucker Canyon to shoe the animal. As he hammered a horseshoe at the forge and anvil, an Indian rode up and dismounted. The squat, square-jawed Apache demanded, "Shoe my horse."

"There's the fire, nails and the hammer. Help yourself," replied Lemuel bluntly. As he turned back to his work, he glanced up to the low bluff behind the ranch house and saw eight to ten mounted Apaches, each holding a rifle. Then and there he decided that discretion was the better part of valor, and shod the Indian's horse.

Again mounted, the Apache said, "Thank you. I am Geronimo." Signaling to his companions, he rode off in the direction of the Mexican border.

The following year this same Indian terrorized the region. After a prolonged *tizwin* (Apache home brew) binge, Geronimo and Nachez, a son of Cochise, slipped off the San Carlos Reservation in what would became one of the West's bloodiest forays. These Apaches and 124 of their followers traveled south along the Arizona and New Mexico line, killing people as they went, until they reached Stein's Pass. From there they turned west, crossed the San Simon Valley and disappeared into the Chiricahua Mountains. They emerged from the other side of the mountains to attack Riggs' ranch in Pinery Canyon, wounding a woman before being driven off.

They next visited the Sulphur Springs ranch of the Chiricahua Cattle Company. Although cowboys at the ranch had received prior warning that Indians were about and had secured the horse herd in the corral, the Apaches came in the night and cunningly emptied the corral. Crossing the Sulphur Springs Valley, they attacked Mike Noonan's ranch and left its owner dead in a doorway, with a bullet hole in the back of his head. This sanguinary foray, which ultimately cost the lives of more than seventy people, left ranchers of the Sulphur Springs Valley shaken and cautious.

The Erie Cattle Company, like other large ranches of the valley, employed enough men to protect its livestock should Apaches cross its range. Fortunately Geronimo did not cross Erie range during his last raid. Instead he headed into Mexico, pursued by United States troops, and after a long chase finally surrendered himself and his followers to General Nelson A. Miles, who brought him back to Arizona. As prisoners the Apaches were loaded into freight cars at Bowie and taken to Florida.

The menace of Geronimo may have been removed, but personnel of the Erie Cattle Company had their brushes with Apaches. On one occasion Lemuel Shattuck and several other cowboys were riding to the McNair place, near Bisbee, to gather a large number of horses. Lemuel spotted what appeared to be an Indian raiding party. The cowboys reined up and while the others watched the Apaches, Lemuel circled around Dixie Canyon to deliver the warning to Bisbee. He then trotted back to McNair's ranch. There he and his companions had supper and spent the night; the Apaches camped close by.

Lemuel slept peacefully in his bedroll all night while the other cowboys sat up, apprehensively watching the Apache campfires a few hundred yards away. Next morning the cowboys gathered the horses, went on their way, and the Indians rode off toward Mexico.

Another time Lemuel and a cowhand, while afoot near Silver Creek, were fired upon by twenty or more Indians. The Indians fired about thirty shots as the cowboys scrambled to a hilltop refuge where they found shelter behind large boulders. Though Lemuel always minimized the danger inherent in such situations, he also stated he was lucky to have survived. There was a little more than just luck involved, however.

In 1934 Lemuel related to Walter Zipf, a reporter for the Bisbee *Daily Review*, that "our brushes with the Indians were few, due partly to luck, but for the most part to foresight and careful planning and being alert at all times. Both cowboys and Apaches used the same trail and sign system in order to keep track of one another. Now and then, the two sides would confront one another, but not often. The cowboys for the most part played a game of spying for their own protection on a hundred mile front, and unless attacked, minded their own cow business. Through their ability to note the movements of the Apaches, they were able not only to guard their herds on the rangelands and in the mountains, but also to drive vast herds along safe trails without danger of attack."

Although Lemuel Shattuck briefly pursued Apaches with an irregular company of Tombstone men raised by rancher William C. Greene,[13] who later became the Copper King of Cananea, Sonora, Lemuel did not have the same fear and hatred of Indians as many of his contemporaries. Perhaps that is because as a newcomer to Arizona Territory he had not seen the atrocities of frontier warfare. Then again his father's Universalist beliefs helped him to tolerate all men and their differences.

No doubt there was a degree of excitement to cowboying. Nevertheless it was hard, backbreaking work, with long periods of tedium, broken only by an occasional frolic in Bisbee, Tombstone, Charleston, or Galeyville. Like his fellow cowpunchers, on payday Lemuel went to town, usually to Bisbee, a copper mining town in the Mule Mountains, or to Tombstone twenty-eight miles to the east. He much preferred booming Tombstone to the dull copper camp of Bisbee, still in its infancy in 1884.

Six years earlier another Pennsylvanian, Edward L. Schieffelin, expressed his intention to prospect the Apache-infested Mule Mountains. His friends thought the expedition foolish and hazardous and suggested that he take his tombstone with him. But courage and enterprise were rewarded. Schieffelin found a rich vein of silver, which he laid claim to as the "Toughnut"; and he christened the district "Tombstone." Ed pulled in his brother Albert and Richard Gird, a knowledgeable mining man from Tucson, and together the men staked additional claims and began development work.

The rich, extensive ore bodies attracted widespread attention; claims developed into mines, and Tombstone became one of the country's richest mining districts. Within four years of the initial discovery the hills surrounding the town were dotted with hoisting works, open cuts, and prospects. Some failed, others succeeded. Of the latter, the Contention, Grand Central, West Side, Head

Center, and the Empire competed with the Toughnut. Along the San Pedro River, eight to ten miles away, mills were erected to process the ore, giving birth to Charleston and Millville. The Tombstone that Lemuel Shattuck first saw was no longer a mining camp; it was a mature district with a population of ten thousand.

Bonanza Tombstone fascinated him. He found it hard to believe the large sums of money that changed hands across gambling tables. He watched faro, dice, poker, blackjack, and roulette. All were new to Lemuel and he tried each one. He quickly learned, however, that a player stood a greater chance of winning at blackjack; it was the one game that intrigued him. He drank a little, too, trying absinthe, a green-colored, aromatic liqueur made from wormwood and other herbs. While he liked the strong, licorice flavor, its sixty-eight percent alcohol content was paralyzing, so he settled for bourbon, less susceptible to doctoring by crooked barkeeps.

Lemuel was always close-mouthed about his times on the town, but it is safe to say that the freewheeling life of a mining town was unlike anything he had known previously. It was masculine, youthful, bawdy, loaded with a sense of opportunity; and there were women — in two varieties. The first group included the so-called good women, few in number, who consisted of maiden ladies, wives, and widows, most of whom ran boardinghouses. For the most part, they were highly respected by miners, cowboys, and gamblers. But the north side of Allen Street was the realm of the second group: dance hall girls and prostitutes, in flamboyant dresses, who catered to masculine needs. They far outnumbered the respectable women, which made Lemuel wonder if the disparity in numbers was due to virtue or merely a quirk of mathematical odds. He laughed at the possible explanations his sweet, gentle mother and strict religious father would offer if they could see this booming mining town.

Tombstone's mineral wealth and the men who developed it intrigued Lemuel. The discoverers of the district had sold their interests in their mines and mills and retired wealthy men. It appeared that others would do the same. The entrepreneurs, drawn to the camp by the scent of boom, looked prosperous. The operators of livery stables, mercantile outlets, lumberyards, restaurants, bars, theaters, and freighters lived lavishly. Hardrock miners and experienced mill workers made three to four dollars a day — a princely sum compared to the $15 to $30 a month a cowboy made. He pondered the two extremes.

Mining camp observations brought home to Lemuel the view that cowboying was an unrewarding occupation; and later reminiscences would indicate that he early recognized the loneliness of life on the range. The quarters to which he was assigned gave him ample opportunity to reflect on that subject. Lemuel lived in a one-room adobe dugout, one-third of its height below the earth's surface, which kept the dwelling cool in summer and warm in winter. He prepared his own meals, no more than cowboy fare of sourdough biscuits, pinto beans, bacon, and coffee. At times he had dried beef, or jerky, which was quite palatable, for it yielded a good gravy to pour over biscuits.

He broke his loneliness by playing cards with other cowboys and by reading, and he had a pet of sorts that he acquired one night. According to a tale he related many times to his children, one night while lying in his bunk, he was astonished to see a little blue racer wriggle out of a hole in the adobe wall. Lemuel lay quietly and watched. To his surprise, the snake settled down beside him. Morning came and his bedfellow vanished into the wall.

During work the next day, he could not drive the incident from his mind, and wondered if his nocturnal companion would reappear. That evening, to his joy, it did. The visits became a nightly occurrence, and before long the serpent would appear as soon as it sensed Lemuel's entrance to the dugout. Life seemed less lonely with this little pet keeping him company, and each evening he hastened back to his reptile friend in the adobe.

Older brother Aus was also lonely and bought a house in Tombstone. His thoughts, however, turned to a young lady, Anna Belle Bootes,[14] whom he had left behind in Pennsylvania. At last he decided to return, marry her and bring her west. After the wedding they embarked on the long trip to Arizona, journeying most of the way by train, and stopping off long enough to enjoy the

Born and raised in Waterford, Pennsylvania, Anna Belle Bootes married Enoch Shattuck in 1885, after a courtship of several years. She spent the first year of her marriage on a ranch in the Sulphur Springs Valley. In 1886 she and her husband moved to Tombstone, where they lived for five years. (Photo courtesy Dan Shattuck)

great New Orleans Exhibition of 1885. The last part of the journey, made by stage from Benson to Tombstone, was hot, dusty, and uncomfortable.

"My experience from Fairbank to Tombstone was a novel one," she commented in a letter to her Erie relatives. "Our conveyance was an old stagecoach drawn by six pretty horses. You would be amazed by the capacity of such an affair. We were all sort of put in by layers, like you pack fruit, till the space is full, and the door was closed. Two persons were on top with all of the baggage travellers usually have. I could occasionally catch glimpses of huge piles of rocks, which Mother Nature has so extravagantly decorated this country with. But then the delightful climate atones in a great measure for the lack of verdure."

Shortly after her arrival in Tombstone, Anna Belle persuaded her husband to drive her out to the range to see young Lemuel. Arriving at his headquarters, she hastened into the dugout. The little racer must have thought its usual companion and friend had returned home, for it came out of the hole in the wall. Anna Belle screamed and her gallant bridegroom rushed in to save her, drew his pistol and shot the snake.

Receiving word he had guests, Lemuel hurried home to see who they were, delighted to have company. He greeted his relatives cordially and looking beyond Anna Belle, spied his little friend lying limp, its head blown to pieces. He could barely contain his grief and asked what had happened. Aus explained that his wife was afraid of snakes so he had shot it. How Lemuel managed to remain hospitable under such circumstances is impossible to guess. But he confessed years later that he never felt quite the same again toward his sister-in-law.

Within a few years Lemuel became restless and left the Erie Cattle Company. Reasons behind his departure are not fully known, for Shattuck was a man who talked little of his past, and seldom reminisced in correspondence. Loneliness and tedium may have driven him from the Erie. More likely, it was vague inadequacy, a feeling common to youngest sons and youngest brothers, that caused his departure.

In Shattuck fashion, Enoch and Jonas were quiet, unassuming men. Leather tough and ramrod straight, the half-brothers had trailed cattle from Texas to Indian Territory. In a young boy's eyes they were the epitome of cowboy life—a tough act for the youngest half-brother to follow. Lemuel may have left the Erie Cattle Company because he felt in some way overshadowed by his brothers, and had to prove himself on turf of his own choosing.

Departing the Erie, Lemuel drifted about southeastern Arizona, working for other outfits. He was employed during roundups by Ernest Bertram Mason, a lower Sulphur Springs Valley rancher who was developing Bisbee's first waterworks, as well as William Lutley, an Englishman who freighted between mining camps and ranched on the west side of the Chiricahua Mountains, close to Erie holdings. Shattuck also helped out at a small ranch in the San Simon Valley, owned by a thirty-year-old Irishman named John Keating. And he became acquainted with the Overlock brothers— Charles, Lemuel, and Will—who ranched near Soldier's Hole in the Sulphur Springs Valley, and operated a butcher shop in Tombstone.[15] Perhaps it was during visits to the latter camp that Lemuel met Constantine Martinelli, from the Italian canton of Switzerland. Although listed as a laborer in the 1884 Great Register of Cochise County, Martinelli, who preferred to be called Joe, was fast becoming a professional gambler.

Another sidekick of Lemuel was a man named Tony Quinn, who, according to Joe Chisholm,[16] developed a flamboyant reputation. While working in Sonora, Quinn and other cowboys, including Lemuel Shattuck, had hazed a herd of cattle up to the cattle pens at La Morita, twenty miles south of the border where customs was located at the time. Getting into a discussion over bullfighting with some Mexicans, Quinn said the sport in Mexico did not amount to a "hoop in hell" and that he and his buddies could give a bull twice a run as a matador. He made his point by setting the rules. "No swords, bandilleros or horses would be allowed." The cowboys would perform weaponless in the corrals of La Morita.

In anticipation of seeing cowboys fight a bull, betting took place on both sides of the border, with about a hundred spectators attending the event. Money was collected as in a modern-day football lottery, to be awarded to the winning cowboy who retrieved a bandanna placed on the bull's horns. Quinn was both master of ceremonies and contestant.

When he confronted the bull, it rushed past him, but he managed to grab its tail, hanging on as the bull raced fiercely around the corral. At last he was able to leap on its back, but soon fell off. The bull reeled and charged him, but Quinn ran and climbed to safety on the fence. Meanwhile, two other cowboys leapt in succession upon the animal's back, seeking the elusive bandanna, but without success. Other cowboys, including Lemuel, tried to no avail until the exasperated animal broke the fence and escaped. Quinn's bullfight was the talk of the border long after he quit cowboying to become a masterful dealer of faro in Daddy Walsh's saloon in Bisbee.[16] Although most of these men were much older than Shattuck, a strong bond would develop between them that would link them in business, and in John Keating's case, in both business and family.

While camaraderie, fun and frolic existed on the range, there was little about cowboy life that inspired Lemuel. It was a hard life, lived out by tough men, who received rough treatment and only coarse fare. It was a narrow life. Indeed it was a rut that prevented a man from seeking a broader existence and achievement of higher ideals. At thirty dollars a month, the best a cowboy could dream of was a small spread of his own some day with a few head of cattle. To Lemuel Shattuck there was more to life, and a quicker road to realization of a man's dreams.

Bisbee in 1882. At upper right is the Copper Queen Glory Hole and buildings housing mine power plant and hoisting facilities. Smokestack right-of-center is that of the Copper Queen smelter. (Fathauer collection)

TWO

A BOOMER

IN THE mind of Lemuel Shattuck, the shortest course to the realization of his dreams lay in mining. At twenty-one years of age, fresh from the Erie Cattle Company's range, much of his knowledge of that profession was based on hearsay, conditioned by a good amount of fantasy. As he had proved to his half-brothers, he was a quick learner, and range life had given him strength and stamina. There was no doubt in Lemuel's mind that he could master the complexities of the mining profession. Just where and how to start may have perplexed him.

The Tombstone he beheld in 1883 was a city of 10,000 population, its mines and mills operating around the clock. The town oozed with prosperity — prosperity propped up by monetary policies of the United States government. Silver had a checkered role in the country's monetary system. In 1834 Congress set the ratio of white metal to gold at sixteen to one, but issuance of paper money during the Civil War drastically disturbed that ratio. Although silver recovered when paper money was gradually retired from circulation after that conflict, its rebound was short-lived: in 1873 Congress dropped the silver dollar from circulation (except for export purposes). Western mining interests proclaimed the action the "Crime of '73," and brought sufficient pressure to bear to secure passage of the Bland-Allison Act in February 1878, which provided for monthly purchase of two to four million dollars in silver bullion, to be minted into dollars at a ratio with gold of sixteen to one. Thus the price of silver was pegged at approximately $1.29 per ounce, a boon to Western mining. The law of supply and demand, however, created havoc.[1]

A torrent of silver flowed from Nevada, Colorado, and Utah, eroding the price of the metal. By the mid-1880s silver had declined below a dollar an ounce, forcing closure of marginal mines. Those mines in Tombstone that did not close cut wages 25 percent, resulting in a four-month-long strike that crippled business and forced many people from the town. Although the strike was settled, the price of silver continued to decline, and one operation after another shut down. Milling at Charleston ceased, and in May 1886 the concentrating works were dismantled and moved elsewhere.

And the calamities continued. On May 12, 1886, a fire destroyed the hoisting works and pumps of the Grand Central Mine. Had there been mutual pooling of resources among the district's mines Tombstone could have rebounded from that disaster. The pumps of the Contention were large enough to have held the flow of water in check until the pumps of the Grand Central were restored. But the companies bickered over costs and refused to cooperate with one another, while the water rose. A fire at the Contention forced cessation of pumping, and sealed the fate of Tombstone.

While he may not have fully understood all the reasons behind Tombstone's collapse, its decline was more than apparent to Lemuel Shattuck in 1887. Bisbee, twenty-five miles to the southeast, offered more opportunity. Although the price of copper was low, the metal had not declined to a point that forced closure of Bisbee's mines, and he stepped into mining as an apprentice at the Copper Prince smelter, one of three refineries processing ore from the camp.

The Copper Prince Mine was anything but impressive. Begun in 1882, it was a shallow shaft and tunnel just north of the Copper Queen Glory Hole, and its smelter was identical to that of the latter mine: a shed containing two 36-inch water-jacketed furnaces manufactured by the Rankin & Brayton Company of San Francisco.[2]

For ten hours a day, six days a week, Lemuel worked at a variety of jobs: pushing ore through the Blake crushers, shoveling coke into the roaring, belching furnaces, and weighing ingots of copper matte. It was hot, backbreaking labor, but it paid better than cowboying.

Lemuel found the contrast between Bisbee and Tombstone astonishing. The copper camp was no metropolis of 10,000 souls — its population in 1887 was no more than 1200, housed in tents, wood frame shacks and adobe structures at the intersection of two defiles of the Mule Mountains: Mule Canyon and Brewery Gulch.[3] For the most part, Bisbeeites were hard-working miners and millmen, drawn from mining districts throughout America. The majority were of Anglo descent — Irish, Welsh and Cornish, the latter dubbed "Cousin Jacks." They were highly productive, proud of their skills, loyal to the companies they worked for. The camp also had a hard core of Hispanics who performed menial chores, such as chopping and transporting wood, burning charcoal for the smelters, carting water, and performing light construction. They were never permitted underground; and Asians, who supplied the camp with vegetables, were banned entirely from town at nightfall. Bisbee in 1887 was a "white man's mining camp," and would remain so for decades.

There were bars and dancehalls, and a section of cribs, but mining companies and townsfolk tolerated no lawlessness. One of the few amenities in Bisbee was the library, founded by James Douglas, the Canadian-born mining engineer in charge of developing the Copper Queen Mine. Confronted with a bloated corpse of a Mexican troublemaker dangling from a tree, Douglas felt a few books might serve to smooth the rough edges of the mining camp. Although life in Bisbee was dull compared to Tombstone, it afforded Lemuel the time and opportunity to delve into the mysteries of geology and mineralogy. The earthy copper oxides and carbonates that he handled every day were beautiful. Although these blue and green minerals were the lifeblood of the camp, he hated to see the finely crystallized specimens go to the crusher prior to smelting. He learned in short order to recognize principal copper and lead minerals. When not working, Lemuel read and studied all that was available at the Copper Queen Library pertaining to geology, listened keenly to the advice of expert miners, and noted the prevalent horizontal limestone strata and the brownish-colored gossans that could hide underground riches. He fantasized that perhaps a bonanza awaited him.

Lemuel's employment, however, would be short-lived. In 1885 the Copper Prince had followed its ore body into Queen property, and the latter company filed suit. Litigation of extra-lateral (or apex) rights was resolved by merger of the two companies shortly after he procured work, and the Prince smelter was shut down.[4]

Shattuck was out of a job; conditions at Bisbee were described "as dull," a polite way of saying there was no work for unskilled labor. Always frugal, Lemuel had saved enough money to get himself to another mining district and another job. Perhaps he would do a little prospecting too. He bought supplies and a bedroll and set out afoot to seek the proverbial "pot of gold." In short, Shattuck joined that migratory population known as Boomers, who drifted from one mining town to another. That day — May 3, 1887 — would be etched in his memory.

Bedroll on his back, Lemuel Shattuck left Bisbee by way of Mule Canyon, gazing at the large trees on either side of the narrow road, actually no more than a creek bed. The hillsides beyond were denuded of trees a few years earlier to feed the fires of the smelters. About 2:30 in the afternoon, as he entered the Toll Road, at the entrance to the divide opposite Old Man Banning's one-room saloon, the earth began to shake and rumble. The north-south oscillation frightened him, and he rushed into the structure. There were a few men standing at the bar, all as bewildered as he. The tremor caused most of the bottles to topple off the shelf, crashing to the floor. Terrified, the proprietor decided the world was coming to an end, placed the few unbroken bottles on the bar and invited his customers to drink up heartily. With that the adobe walls crumbled and, though the world did not end, poor Banning was left with a ruined edifice and all his alcoholic stock consumed or destroyed.

The earthquake of May 3, 1887, caused great damage and loss of life south of the border in Mexico. Two hundred people were killed in Bavispe and Arispe, Sonora. North of the international line, the quake leveled most of the adobe structures in Tombstone and Charleston, and several in Bisbee; damage was reported as far away as Tucson and El Paso.[5]

Shakened but unscathed, Lemuel gathered up his belongings and continued on his way. The road at the bottom of the divide was covered with boulders, some massive in size, that had tumbled down the hillsides during the quake.

Lemuel followed the San Pedro River northward, exploring likely looking gossans or signs of mineralization in the hills along his route. Out of necessity, he paused now and then to work at any available job. Near present-day San Manuel in Pinal County, he toiled in the Mammoth Mine, mucking and tramming gold ore, as well as driving a jerk-line mule team hauling ore from the mine to the stamp mill located a few miles away on the San Pedro River. Wanderlust drove him on to the Pioneer Mine, located on the west side of the Pinal Mountains above Silver Creek in Gila County. Eventually he arrived in the Globe District, where he mucked and mined copper in the Old Dominion, and placer mined in Lost Gulch, which later became part of the Northern Inspiration Company.[6]

Lemuel skipped from Globe to the placer fields of Lynx Creek, Walker Gulch, and Black Canyon, and eventually ended up cleaning out old stopes at the Vulture Mine, a famous gold mine fifteen miles south of Wickenburg.[7] Shattuck stayed only long enough to save enough money to push on to another mining locality.

The desire to see the famed silver camps of Nevada drove Shattuck northwestward. Due to the depressed price of silver most of the Washoe mines were closed or working with only skeleton crews. The reduction works of Austin, which had made history processing complex silver ores full of arsenic and tellurides, fascinated him. He would have stayed if there had been work. Crossing the Virgin River, he headed into Utah, stopping at Silver Reef northeast of Saint George, a mining district famous for its unique association of silver with sandstone. Silver Reef had passed its zenith, however; few if any mines were operating and much of its population had moved on to other locales. He pushed on to Cedar City and Parowan, where he observed Mormons mining and smelting iron. He admired the Latter-Day Saints, as Mormons call themselves, for their Christian ethics and industrious habits. And the Bishop of Parowan liked Shattuck, inviting him to join the community. He even offered Lemuel his eldest daughter in marriage. Lemuel never revealed whether it was this offer or the lure of mining that caused him to push on. Nevertheless, he was soon on the move again, investigating mining districts along the western flanks of the Wasatch and Oquirrh mountains. Silver camps universally slumbered. Those that demonstrated some life were being held together by production of lode or placer gold. Everywhere miners talked of the coming day of copper, and Lemuel headed toward Butte, Montana, an old silver district with a copper lining. There he worked through winter of 1887-88.

In spring of 1888 he retraced his steps southward, lingering in Globe to placer mine with a partner. He invested everything in the venture, with some success. In the hope of selling their claim, they placed the gold in a glass jar to tempt prospective buyers. One day, Lemuel's partner went into Globe, saying he would return later. No partner appeared at day's end, and Shattuck went into town, stopping at a saloon for a drink. On a shelf behind the bar sat a familiar glass jar containing gold flakes. Suspicious, he inquired as to where it had come from. The bartender replied that earlier in the day a man had come into the bar, saying he needed cash as he had to leave town. Disillusioned, Lemuel swore never to be duped again, packed his belongings, and struck southward for Cochise County, the Mule Mountains, and Bisbee.

Two years of wandering from one mining camp to another had killed the appeal of a prospector's life. There was too much uncertainty attached to full-time prospecting for his liking. His sojourn had not been wasted, however. Lemuel Shattuck had seen the "Elephant"; he knew men could make money at mining, not from silver perhaps, but from other metals. For the moment the cry was for gold, but everyone anticipated that copper would skyrocket.

The building trades and electrical engineering were increasing demand for the red metal.

He was footsore and weary when he crossed the flats and entered the divide. Pausing to rest, he looked down upon the familiar camp. The town had changed during the two years he had been gone. The old Prince and Copper Queen smelters, the latter situated at the foot of the Glory Hole on the mountainside, were gone. A giant new reduction works stood on a bench just north of the Czar Shaft. The foundations of the smelter were being poured when he walked out of Bisbee in 1887. In 1888 four water-jacketed furnaces had been blown-in, and now a pall of yellow-gray sulphurous smoke filled the canyons of the Mule Mountains. Increased mining activity was apparent: there were several new workings. The Goddard shaft, sinking of which had started just prior to Lemuel's leaving town, was completed. Three other shafts were being sunk: the Holbrook, Spray, and Gardner. Increased mining activity had brought additional population, and the town seemed more substantial. It had actually crept further up the canyons. Durable frame buildings, some with several stories, had replaced makeshift shacks and tents. Most important, railroad tracks had been laid into town. In a never-ending battle to reduce freight rates, the Copper Queen Consolidated Mining Company had linked the Mule Mountains to the outside world in 1888-89 by building the Arizona and Southeastern Railroad to Fairbank, where it linked with the New Mexico and Arizona Railroad that ran from Benson to Nogales.[8] Bisbee had indeed changed during his three year absence. The thought that perhaps his future lay in these mountains fluttered across his mind. Thus began Lemuel Shattuck's love affair with Bisbee, which would last a lifetime. As years passed, that affection would embrace all of Arizona.

Lemuel Shattuck walked into Bisbee flat broke. He possessed one shirt, the one on his back. He was long-haired, unshaven, covered with the grime of the trail. A man walked up to him and said, "Young fellow, you look beat and down on your luck," and handed him a $10 gold piece. Lemuel thanked his benefactor and headed toward Brewery Gulch to seek food, a bath and shave, and lodging. In the morning, he would find a job.

He trudged up Brewery Gulch, obtained something to eat at a restaurant, then searched for a room, finally finding accommodations at a boardinghouse close to the "Line," paying the rent with the ten dollar gold piece. With the thought that he would seek employment in the morning, Lemuel retired early. Arising before dawn, he had breakfast, and before start of the day's shift, made his way to the general offices of the Copper Queen Mining Company, where he signed on as a trammer at the Czar shaft, a position that paid $3.00 a day.

His entrance into Bisbee had not gone unnoticed. Isabella, eldest daughter of Edward Grenfell, a Cousin Jack shift boss at the Copper Queen, observed Shattuck's passage up Brewery Gulch from the window of the family home, built on the hillside of Brewery Gulch.

"Who is that unkempt young man?" she asked her father, sitting beside her.

The old miner, who had seen hundreds of drifting miners before, curtly answered, "He must have just come to town, probably another good for nothing ne'er do well."

Isabella Grenfell said no more, but she could not dismiss the young man from her mind. There was something about the man that appealed to her. Perhaps his stature and bearing, or the glint of his eyes. Intuitively she knew he was not the average tramp miner who wintered in Bisbee and drifted northward in summer. Yet she suspected that her haughty mother would disapprove of this new arrival, just as her father had done — the Cornish looked down on all that was not Cornish. And Edward Grenfell was very much a Cornishman.

Edward Grenfell was one of thousands of Cousin Jacks who immigrated to the United States and applied their skills and knowledge to American mining. "Peerless miners" they were called, both for the brand of tobacco they smoked and their ability underground. Edward was born in 1845 in the tin and copper mining center of Truro, Cornwall, the son of a minister who was killed while trying to separate two men in a drunken brawl. Edward came to the United States as a teenager, finding employment in the lead mines at Suscasunna, New Jersey. There he married Elizabeth Paul and established a family which grew rapidly, eventually numbering twelve children. Isabella (christened Elizabeth Isabella) was the eldest daughter, and was called Belle by her family.

Like all miners, Edward drifted from mining district to mining district. He worked in the copper mines of northern Michigan, and then made his way to Leadville, Colorado. Tombstone was his next stop. When decline of silver closed that camp's mines, he packed up his family and moved to Bisbee, obtaining a shift boss position with the Copper Queen Consolidated Mining Company.

Isabella found the drifting life of professional miners distasteful. She disliked the hardships of mining camps, and never adjusted to Arizona, preferring the more civilized East and Midwest, where seasons stood in sharp contrast to one another. Belle, as her father preferred to call her, was an intelligent, educated, well-read, and attractive young lady. At age eighteen, she was hired to teach school at Sunnyside, a religious community founded by Sam Donnelly on the southern side of the Huachuca Mountains near the Copper Glance Mine. In fact, it was at that community that she experienced the earthquake of 1887. Just as it had been for Lemuel Shattuck, that day would be monumental for Isabella.

At Sunnyside Isabella was separated from her family for the first time. The young teacher boarded with a family, the patriarch of which imbibed freely, often becoming intoxicated, unruly, and abusive. At such times, Isabella barricaded herself in her room, fearful of a replay of the situation that had killed her grandfather. The day of the quake the landlord and his cronies reached their usual state of intoxication. The man's wife became so exasperated with his drunkenness that she took him outside and tied him to the wheel of a buckboard. Then she ordered his companions out of the house. Shortly thereafter, the earth quaked and a large fissure opened in front of her husband. Doomsday had come with clouds of dust and fracturing of the earth — it was indeed a sobering moment, and the man cried in repentance. He never touched alcohol

Elizabeth Isabella Grenfell, eldest daughter of Edward Grenfell, married Lemuel C. Shattuck on April 13, 1891. A loving wife, and a doting mother, Belle was Lemuel's helpmate for thirty-three years.

(Fathauer collection photo)

again, to the relief of his wife and boarder. In 1888 Isabella returned to the family home in Bisbee.

Isabella aided her mother in sewing and mending clothes for a family of industrious miners, sitting in a favorite chair by the front window, from which she viewed the endless comings and goings below in the Gulch: Mexicans prodding their burros laden with bundles of wood cut from surrounding hillsides, miners coming off shift swinging empty lunch pails, on their way to boardinghouses and modest cottages. She often observed "Yellow Dog" Brown, followed by his golden colored whippets — his constant companions — on the way to his saloon and gambling establishment in the lower Gulch; horse-drawn carts filled the defile, their drivers taking grocery orders in the mornings and delivering them in the early afternoons; children went by on their way to and from school, as did occasional ladies of the night from "the Line." Everyone walked, as all businesses were concentrated in the lower areas of two canyons. Only a few affluent citizens could afford a horse or carriage. Both could easily be rented at the O.K. Stable located at the junction of Main Street and Brewery Gulch. And it was from that same window that Isabella Grenfell kept track of the comings and goings of Lemuel Shattuck.

Lemuel frequently observed the chestnut-haired beauty staring down at him. A certain sadness, or was it haughtiness, about her was enticing, and one day Lemuel stopped and greeted Miss Grenfell and she answered him. Their conversation gave him the chance to observe her more closely. He liked what he saw: fair complexion, expressive eyes, a straight nose. He estimated her height at a little over five feet. She was not slender, but pleasingly rounded in all the right places, and well-corseted. It was the beginning of a courtship that, according to family tradition, was carried on through a window.

Isabella observed that this young miner was popular with Bisbeeites, always having a friendly greeting for everyone. He was tall, muscular, blue-eyed. It was not long before she eagerly awaited his passing. Each time they talked she discovered a little more about him: he was a Pennsylvanian who had come West to join his brothers' cattle business; he was beginning to speak Spanish, and she was surprised to learn that he could speak German as well. He was not of the usual run of miners.

Each day Lemuel lingered longer at the window, their conversations becoming increasingly more personal. Suddenly one day, he asked her to marry him. She responded with an immediate "no." The rebuff did not discourage him and daily he pressed his suit, always receiving the same negative answer.

While she liked this gentle miner, Isabella feared the disapproval of her parents. Moreover, she did not want a life like that of her parents, spent in isolated, rugged mining camps. In Succasunna, New Jersey, her older brother, a mere toddler, had succumbed. In Michigan, while walking a shortcut to school through the woods, she had come face to face with a bear. In Colorado, three or four Indians entered the Grenfell home, and traipsed through every room until they found the kitchen. They lifted the covers of pots bubbling on the stove, and thrust their fingers in to both smell and taste the food. The fare apparently did not appeal to them, for they shrugged their shoulders and walked out the kitchen door, much to the relief of her mother and the children.

Isabella especially disliked Arizona mining camps, with their wildlife — scorpions, centipedes, coyotes, snakes, bobcats — and constantly barking dogs. There were gamblers, carousing miners on payday, and raucous cowboys. Occasional shootouts and now and then a public hanging were even worse. The sun beat down day after day. During July and August lightning streaked across the sky, followed by deafening, rumbling thunder and torrential rain. In Bisbee raging floods poured down the two canyons; sulphurous smoke from the smelter always tainted the air. She doubted she could ever accept and like Arizona, or that she wanted to spend the rest of her life tied to a miner; yet she wondered where she would meet someone who would give her a gentler lifestyle.

Despite all her misgivings, Isabella realized she was developing strong feelings for Lemuel and wished she could learn more about him. But she would not have that chance, for the day came when he walked by without glancing up in her direction. He passed the house without stopping to talk. Days went by and he continued to ignore her. She could not believe this was happening to her. Dismayed, she longed for him to extend at least a brief greeting — oh, if only he would propose again!

Lemuel's feigned disregard of Isabella was effective. It forced her to a decision. She waited for him to pass by and, when he did, threw open the window and called, "Lemuel, please stop. I will marry you."

"When?" he asked.

"Today," she answered.

Overjoyed, he rushed down the Gulch to the O.K. Stable and rented a buckboard and horses. When he returned, he found her waiting in the road. He heisted her up beside him and they were soon on their way to seek out the justice of the peace at the courthouse in Tombstone.

Edward Grenfell undoubtedly would have stopped the eloping couple, with force if necessary, had he not tarried longer than usual with his friends in a favorite saloon. When he got home, his daughter was nowhere to be found. Fearing something had happened to her, he walked down Brewery Gulch inquiring if anyone had seen her. Someone volunteered the information that she had been seen in a buckboard driven by Lemuel Shattuck, headed up Tombstone Canyon. Furious, Edward hurried home, saddled his horse and galloped off in pursuit. He arrived in Tombstone too late: Isabella and Lemuel were already married — on April 13, 1891.

His daughter had married a common miner, who was not even Cornish. For the moment Edward Grenfell seethed. The young couple did not return to the Grenfell residence, but purchased a small house above Brewery Gulch, nearby. Isabella was a meticulous housekeeper and a good cook, although she found western fare, which Lemuel relished, unappetizing. She was not a coffee drinker, preferring tea instead. Both detested the potable water brought in on donkey back from springs in upper Brewery Gulch, which was often contaminated and the source of typhoid and other infections. Lemuel drilled his own well, no easy task due to the limestone strata. Water of most Bisbee wells was usually reddish in color, brackish, and metallic tasting. The Shattuck well, however, produced clear drinking water, which the couple shared with neighbors.

Lemuel continued working for the Copper Queen Consolidated Mining Company. Although he disliked tramming, which required great strength and endurance, he did his job to the best of his ability. Fellow workers, particularly Baptiste Caretto,[9] who worked in the same drift, marveled at Shattuck's strength and determination. When Lem broke the arm of a man attempting to rob a saloon owner, he attained the title of strongest man in the Warren District. Will Grenfell, Isabella's brother, was proclaimed the most handsome man in camp.

In time Edward Grenfell's hostility cooled. He found his son-in-law amiable, hard working, and ethical in his dealings with everyone. The Cornishman reluctantly admitted that Lemuel showed a flair for mining. When doubt disappeared that Lemuel would take good care of Belle, they became close friends. The younger Grenfell children also grew fond of the young miner, and looked upon him as an extra protector they could rely upon.

On Isabella's birthday the following year, Henry Spencer Shattuck was born. The parents were delighted with the strong, healthy baby. They doted over the child, and the younger Grenfell children shared the time-consuming task of watching over little Henry. Within the next four years, two more sons were born, Warner and Edward. Bisbee, besieged by epidemic diseases, was not a good place to raise children, and Edward became desperately ill with an ailment doctors diagnosed as spinal meningitis. Despite all attempts to save the toddler, he died, and Isabella was grief stricken. Although Lemuel grieved, he accepted the cycle of man — birth, life, and death — much better than his young wife, and he did his best to console her. Another baby was due in a month or so. But she clung to the belief that the greatest tragedy in life was the loss of a loved child.

Despite Lemuel's hopes for a girl, the baby was a boy. Neither parent could think of a name until an old German miner stopped by to see the baby and suggested the name of Bismark, who shared the same birth date. Half the suggestion was accepted, the parents settling on Mark. He had the Grenfell good looks, and the robust body of a Shattuck. A strong relationship developed between Mark and his mother, perhaps partially due to Edward's death.

The 1890s was a period of rapid growth for Bisbee. In 1888 the Secretan Syndicate, a group of French commodities speculators, attempted to corner the international copper market. For over three years a million pounds a month of Copper Queen metal went to the Paris brokers, at 12½ to 14½ cents a pound.[10] Flushed with capital, the mining company, under direction of James Douglas, a shrewd Canadian-born metallurgist, and brothers Louis and Ben Williams, equally knowledgeable Cornish smeltermen, purchased one claim after another, sunk additional shafts to tap immense lenses of mineralization, built a railroad and a general store. Bisbee blossomed. It was apparent that millions of dollars were being made mining copper, and millions more could be gleaned by supplying the needs of a growing populace. Even before his marriage to Isabella, Lemuel weighed the opportunities at hand, and began to delve into lines of business other than just mining.

Lemuel's earliest commercial endeavors at Bisbee were related to the Erie Cattle Company. In 1890, shortly after his return from his tramp around the West, Lemuel was commissioned by the company to purchase a group of claims where Lowell now stands for stockyards and a slaughterhouse, as well as plots of ground on Bisbee's main street for the location of several butcher shops. Although the Erie Cattle Company retired from the butcher business, turning that enterprise in 1891 over to "three practical butchers"[11] (the Overlock brothers: Charles, Lemuel, and Will), the stockyards next to the tracks of the Arizona and Southeastern Railroad proved a boon not only to the Erie Company, but to other ranchers of the Sulphur Springs Valley. For nearly a decade, the spring and fall roundups ended south of the town. Thousands of head of cattle were penned on Erie property and driven through Erie chutes into waiting railroad cars to be transported to eastern commission houses, and Los Angeles slaughterhouses. Cochise County cattle stocked the ranches of Kern County, California, and were shipped to Kansas for fattening before being moved eastward. Acquisition of those claims in Lowell turned Bisbee into both a copper and cattle town.

To augment his wages as a trammer, Lemuel Shattuck located a plot of adobe earth in upper Brewery Gulch, close to springs from which former Sulphur Springs Valley rancher E. B. Mason drew his water to supply the needs of Bisbee's population. During off hours Shattuck manufactured and sold adobe brick. This enterprise went so well Lemuel hired a crew of Hispanic laborers and contracted to erect and repair buildings. The abode brickyard led Shattuck into a deeper affiliation with E. B. Mason.

Mason was one of Bisbee's earliest residents, arriving in camp in 1881 to take a job with the Hardy Brothers mercantile establishment. Shortly thereafter, he claimed several springs in upper Brewery Gulch and began packing water on muleback to homes and businesses. The venture was an immediate success and the Bisbee Water Company was born. As might be surmised, Mason furnished Shattuck water to make his bricks. The business relationship deepened, and on April 19, 1892, the two men jointly filed a location notice for a reservoir site in Dixie Canyon.[12] Four years later a 45,000-gallon water tank was placed on the site.

How involved Shattuck was in Mason's water company is not fully known; but Lemuel had considerable interest in it. The fact that he had joined in partnership with the proprietor of a commercial water company indicates Shattuck was firm in his belief that Bisbee would have a long and prosperous life.

Early in 1892 Lemuel Shattuck left the Copper Queen Mining Company to devote all his energy to business. Sometime during the first half of that year he acquired property — probably by right of possession — on Upper Brewery Gulch, adjacent to his residence, and erected a lumberyard.[13] Lemuel also moved into town entertainment. In summer of 1893, again in partnership with E. B. Mason, he opened the town's only skating rink,[14] one of few enterprises, except saloons, that catered to the recreational needs of the

The Shattuck Lumber Company on Brewery Gulch in the mid-1890s. Lemuel Shattuck in foreground. At this time the Shattuck residence was the house immediately behind the lumber shed.

(Photo courtesy Bisbee Historical and Mining Museum)

community. Under the management of Will Grenfell and a man named Howe, the rink, according to the Tombstone *Prospector* of January 3, 1896, was "open Wednesday and Saturday evenings. Ladies Free on Friday afternoon from 2 to 4 o'clock." The rink, no more than a wooden floor and orchestra area shielded by a wooden canopy, was utilized for more than just roller skating. Parties, dances, socials, and weddings were booked there. How successful the rink was as a business, no one knows. What is known is that in November 1894 E. B. Mason transferred his half interest in the enterprise to Shattuck;[15] and by end of February 1896 ads for the rink disappeared from the newspapers.

As the skating rink venture suggested, Lemuel was pooling his talents and resources with those of his in-laws and close friends, a practice he would continue throughout his life. In mid-November 1894 he and former rancher John Keating, who had married one of Isabella's younger sisters, bought a lot and house in lower Brewery Gulch from W. L. Humphrey for $65.[16] They renovated the structure, "painting and brightening" it up, installed a long bar with footrest, mirrors, chandeliers, erected a cold storage room, and obtained a franchise from Anheuser-Busch Brewing Association, giving them exclusive distributorship in Bisbee, in competition with Frank Dubacher and Joseph Muheim, who sold Pabst Beer. On January 2, 1896, the Tombstone *Prospector* ran the first ad for the St. Louis Beer Hall — Shattuck & Keating, proprietors. "Ice Cold Beer on Draught."

In comparison to the Opera Club and Free Coinage Saloon, with their mahogany bars and crystal fixtures, the St. Louis Beer Hall, at 12 Brewery Gulch, was a modest, homey establishment. Like all such enterprises in a "man's town," it catered to masculine needs. Bartender Otto Schmidt, impeccably dressed with white apron, dispensed not only alcoholic beverages, but also attention, sympathy, and good conversation. It was a place where men could escape the drudgery of hard, often dangerous, physical labor. Unlike today's saloons, the St. Louis Beer Hall was bright and spacious. There was light enough to read the always available current editions of newspapers and magazines, and all popular brands of tobacco, cigars, and smoking accessories were stocked.

Women were not allowed in the Shattuck and Keating saloon. With drinks in hand or at their elbows, under a cloud of pungent tobacco smoke, men were free to discuss, in their own manner, town news and gossip. If they wished they could indulge in poker, blackjack, or any other game of cards. L. O. Milless, "one of the squarest of gamblers," ran the roulette game, and cool-headed Tom Wilson, the faro bank. Shattuck and Keating were not above sitting in on a game, if invited — both men could hold their own with professional gamblers. In the game of whist, Lemuel's dexter-

The St. Louis Beer Hall sometime between 1894 and 1897. Lemuel Shattuck stands in doorway. To his

ity always amazed the Shattuck family. "Papa could shuffle and deal cards like the most efficient and expert gambler."

The St. Louis Beer Hall, a workingman's bar in the best sense of the word, was an immediate success, for on March 16 the *Prospector* proclaimed "there are few industries here which have risen to such prominence or deserve more favorable notice than that of Shattuck and Keating. They carry an excellent selected stock of liquid refreshments, embracing the leading brands of imported whiskies, brandies, gin, rum, and still and sparkling wines; also a full and complete line of the best brands of cigars. These gentlemen stand high in business circles, and those doing business with them find them both pleasant and sociable as they are always generous and their patrons well received."

As sole distributors of Anheuser-Busch Beer, Shattuck and Keating moved swiftly to build a sizable wholesale business, which they called the Bisbee Beer Bottling Company.[17] When Lemuel and Charles Overlock drilled a well in front of their butcher shop near the St. Louis Beer Hall, Shattuck purchased a 10-horsepower refrigeration plant from St. Louis, and the men erected a cold storage locker to accommodate both the liquor and butcher businesses. They were soon furnishing ice cold bottled beer to other saloons, restaurants, and boardinghouses throughout the Warren District, as well as to clients in Tombstone.[17] Just how large a business Shattuck and Keating had can be ascertained in records of beer shipments coming into Bisbee. The St. Louis Beer Hall was receiving at least a carload of beer a month, much to the delight of thirsty miners. "May they never run out of stock," trumpeted the *Prospector* on June 9, 1896.

While the St. Louis Beer Hall prospered from the very start, Keating felt confined by the saloon. In temperament he was a prospector and mining promoter, and in November 1897 he sold his interest in the business to Lemuel. It was not the end of their friendship, however, for the two would remain close and work together in other enterprises.

The St. Louis Beer Hall was Lemuel's sole concern; what went on there was seldom discussed in the family circle. It was different with the lumberyard—that was a family business, the prime concern of both Lemuel and Isabella. For several years they had almost a monopoly on the supply of building materials. They branched into contracting and construction, and until 1896 their wood went into the majority of frame houses built from Tombstone Canyon to Naco Road. Whenever she could, Isabella crossed the street and lent a hand in managing the business, keeping books and attending to sales and orders, while Lemuel ran the bar and looked after their mining properties. She also managed ten cottages along Brewery Gulch, which the Shattucks' rented to miners. He admired her intelligence and considered her a valued helpmate.

By 1896 construction at Bisbee reached a fever pitch. Construction could not keep pace with population growth. Roads and paths were graded along the hillsides skirting both Tombstone Canyon and Brewery Gulch. Pads were blasted on the hillsides for placement of houses, boardinghouses, and miners' shacks. Commercial buildings were erected along the town's two main thoroughfares. Other entrepreneurs moved in to carve their niche in the building trade, among them long-time Bisbeeite August Geisler, who opened a lumber business in early 1896, and Jack and Chase who had originally started near Naco and later moved to Bisbee. As representatives of the Blinn Lumber Company of Los Angeles, they dealt in Oregon pine and redwood.[18]

Bisbee's mineral wealth lay in soft, shifting formations that required enormous amounts of reinforcement to keep tunnels, drifts, and stopes from collapsing. Almost from its inception, the Shattuck Lumberyard stocked heavy timbers, obtained from the Pacific Northwest, for mine support. This timber, in dimensions of 10 x 10, 10 x 12, and 12 x 12 inches, and lengths up to 12 feet, had to be sawed to precise lengths and mortised to fit into square sets, the universal method of erecting supports within the stopes. Only the district's larger mines had sawmills equipped to handle this lumber. To accommodate smaller mine operators, the Shattucks installed a sawmill at their yard to cut cribbing and frames. Thus Lemuel drew upon his knowledge of mining to sidestep competition in the lumber trade.

For a while at least, there was more than enough construction in the Mule Mountains to satisfy all three proprietors. In one year Bisbee's population jumped by a thousand people. On November 17, 1897, the *Prospector* noted that the town's population numbered 4,000, "not counting our Mexicans. The town continues to extend its lines, and the surrounding hills are now dotted with many pretty dwellings. . . . We have our Nob Hill, but as yet no multimillionaires." In another issue the paper stated "our lumber merchants are in receipt of large consignments of building materials. Bisbee has no boom, but is constantly increasing, elegant and substantial cottages being erected."

The *Prospector* was right. Bisbee had not yet produced any multimillionaires. That would change very quickly. The 1890s saw expansion of mining in the Warren District. The Copper Queen Consolidated Mining Company had extended its property through acquisition and development of claims. Striking of rich copper carbonate in the Spray Mine proved beyond doubt that mineralization was not confined to limestone horizons close to the town, but extended through fractured formations lying further to the southeast. Experienced mining men and knowledgeable prospectors, including Edward Grenfell and Lemuel Shattuck, long surmised that ore deposits radiated outward from the giant porphyry pile known as Sacramento Hill.

Anticipating extension of mining, Lemuel prospected the Warren District and staked what he considered promising mineral occurrences. In 1893 he filed on five claims in a canyon high on Copper Queen Hill, one of which he called the "Shattuck." The following year, in partnership with saloon owner Martin Costello and Pete McCoy, Lemuel staked the Waddell and Top Gallant.

Not only did Shattuck prospect and file for mineral rights, he also purchased them, usually in partnership with others. With Wallace Dwyer, and three men named Green, Clark, and Dean, Shattuck purchased the Julia prospect; and he, Edward Grenfell, and John Keating purchased nine claims on the slopes of Sac-

ramento Hill, about two-and-a-half miles east of the heart of Bisbee and a mile beyond the present-day Evergreen Cemetery; that
included the Pay Day, Comet, Come-by-Chance, and the Primrose. Collectively called the World's Fair Group, these properties
had a history of ownership.

The properties were first located in the early 1880s by Nicholas
McCormick, who commenced exploration but soon ran out of resources and abandoned the property. He was subsequently mur-
dered by Mexicans. Shortly thereafter, Bob Stevens came into possession of the property, did a little work on the claims, and then
sold them to George Hill, who in turn passed them on to Grenfell,
Shattuck, and Keating. In 1895 they began work and over the next
five years developed the property.[19]

They dug an inclined shaft, erected a mule-powered whim, and
burrowed for 400 feet, running drifts at intervals. Pockets of
copper carbonates — azurite and malachite — were struck on the
120-foot level, and sixty feet below that they penetrated an "in-
teresting" cave rich in oxide ore. At the 200-foot level considerable
drifting was done, which revealed "nice carbonates" to the west;
and a drift in the opposite direction showed lean ore all the way. A
respectable body of low grade sulphide ore showed up in a short
crosscut at the 380-foot level; and the bottom of the incline rested
in a formation distinguished by liberal showings of carbonates and
black manganese, the latter indicative of strong mineralization at
depth. By 1900 the property was fully opened, with sizable ore
bodies in sight. Several tons of ore sat on the dump.

Shattuck was untiring in his quest for mining properties. By
1900 he had filed on forty-five claims. Most were in the Warren
District, but others were at the Hartford District in the Huachuca
Mountains and in the Turquoise District, near Gleason.[20] Copper
and gold deposits further afield also caught his attention. In spring
of 1902 he, Peter Johnson, Joseph Muheim and E. G. Austin
dredged up $65,000 to bond the Tenochio Mining property,
twenty-five miles east of Fronteras, Sonora.[21] That same year
Shattuck acquired three placer claims in Idaho,[22] and became in-

terested in a rich gold strike, known as El Tigre, sixty miles below the border in the Moctezuma District of Sonora. The quest for mining properties, and the funds with which to develop them, would ultimately lead Lemuel Shattuck into the world of high finance.

THREE

MURDER IN THE SALOON, AND
THE MINERS AND MERCHANTS BANK

AT TURN of the century Cochise County boasted the largest population of any county in Arizona Territory. Ten thousand persons lived in Bisbee, packed into tiers of closely built houses that climbed higher and higher on the hillsides. A thriving business district mushroomed in the canyons below, and mining development reached into the lower end of the district, creating pockets of population that would grow into satellite communities. This growth created problems. Outhouses built in the canyons polluted the water. Trash and offal were dumped behind boardinghouses. Sanitation was non-existent, fire protection and law enforcement inadequate, and the town had no public utilities. Because they built their residences and businesses on government land, Bisbeeites held no title to their holdings.

In November 1900 the Copper Queen Consolidated Mining Company spearheaded a movement to improve conditions by incorporating the Bisbee Improvement Company. Sitting at the helm of this public service corporation was Walter Douglas, manager of the Copper Queen Mining Company and son of James Douglas. William H. Brophy, head of the Copper Queen Mercantile Company, was vice president; and the position of secretary-treasurer was filled by S. W. French, assistant superintendent of Copper Queen mines. Representing the town's commercial interests were mercantile proprietor John B. Angius, and saloon keeper and lumberman Lemuel C. Shattuck.

Capitalized at $50,000, the Bisbee Improvement Company was incorporated in 1901 and immediately embarked upon a program to develop telephone communications between the copper camp and the emerging smelter town of Douglas, in the Sulphur Springs Valley. The company built an electric plant in South Bisbee, and acquired an ice plant formerly owned by the Copper Queen. Delivery of ice to homes and businesses proved so popular that the corporation installed at Lowell a modern ice plant of thirty tons per day capacity.

The Electric Age had dawned in America and candles and kerosene lamps were giving way to the light bulb. Not to be outdone by other western communities, in 1901 the Improvement Company added a generating plant to its Lowell facility large enough to furnish electric current for municipal needs. Most important, the company drilled wells at Naco and erected a safe potable water system. A giant step had been taken toward furnishing the town public utilities; only incorporation of Bisbee could assure land titles, fire and police protection, and improve sanitation.[1]

Incorporation of Bisbee was first voiced during summer of 1901. Reception of the plan was at best lukewarm, with many residents, particularly Brewery Gulch businessmen such as Letson, Muheim, Schmid, and Shattuck, openly opposed to it. They suspected the movement to be no more than "a graft of Main Street business men to make poor people pay the expense of fixing that thoroughfare," or to raise money to flume flood waters that swept through Tombstone Canyon during summer months.[2] The practicality of the matter, however, won out, and within a few months these men became the loudest boosters of incorporation. At a mass meeting in October a committee was elected, consisting of I. W. Wallace, J. H. Scott, J. B. Angius, G. H. Willcox, and H. C. Stillman, to fix town boundaries.[3] A petition was circulated to determine the number of Bisbeeites in favor of the proposition.

The boundaries of the new city included Brewery Gulch, Quality Hill, School Hill, Tombstone Canyon and Chihuahua Hill. Attorney J. M. O'Connell placed the petitions and plats before Cochise County supervisors, who entered an order for incorporating the town of Bisbee on January 9, 1902. A temporary city council was appointed consisting of T. M. Shearer, J. B. Angius, Ed Scott, James Letson, Otto Geishenhofer, Peter Johnson, and Lemuel Shattuck. The Bisbee *Daily Review* noted that all were substantial citizens and wished them good luck.

Few of the councilmen had experience in county government or city management. When asked about the merits of incorporation, most ducked the question altogether or evaded probing reporters. Lemuel Shattuck, who had served on the Cochise County Board of Supervisors in 1894-95, as well as having been a delegate to County Democratic conventions as early as 1892,[4] knew precisely what to say: "We want to do everything we can to benefit everybody." Such was the aim of the council when they convened their first meeting on January 21, in a tiny room at the back of the Bisbee firehouse on Main Street, to appoint town officers. Joseph J. Muirhead was chosen mayor; I. W. Wallace, secretary; Dayton Graham, town marshal; and J. M. O'Connell, city attorney.

Over the course of the next sixty days resolutions were drafted and passed fixing taxation of various businesses, and providing for fire and police protection. The first code, passed on January 22, spelled trouble for Brewery Gulch: "It shall be unlawful for any women within the limits of the Town of Bisbee, in any saloon or in any room or apartment adjacent to such saloon or connected therewith either to hire or otherwise, to sing, dance, recite, play on any musical instrument, give any theatrical performance or other exhibition, serve as a waitress or bar maid, or engage or take part, either as employee or otherwise, in any game of chance or amusement played in any saloon or room or apartment adjacent thereto . . ." Violation of this ordinance, designed to maintain the saloon as a man's world, carried a stiff fine of not less than $20 nor more than $50 and/or ten to twenty-five days in jail.[5]

This ordinance was bitterly contested by some owners of saloons and gaming parlors, who felt that women on their premises were good for business. Their fight, however, was futile, for Bisbee was experiencing the first swell of a wave of morality generated by the management of the Copper Queen Consolidated Mining Company, which desired a mature image for the town. A "wide-open" town catering to the needs of transient miners would not be tolerated. Bisbee was there to stay and what was needed was a stable aggregate of miners with families, as well as professional and business people who would leave an indelible mark on the community.

To raise revenue, the council created a licensing system for all forms of enterprise. In July a monthly fee of thirty dollars was levied against "every proprietor of a bawdy house, or house of prostitution within the city." Police raids of whorehouses and open gambling followed and by spring 1903 the town's population of "lewd" women had been confined to the upper end of Brewery Gulch. "Bisbee has outgrown the wide-open stage," proclaimed the *Daily Review*, "and is now a city that recognizes law, order, and public decency."

Despite the council's attempt to mold Bisbee's image, it was still a town where nearly every man carried a gun, and where there were few diversions for unmarried males except the saloon, gaming table, and the prostitute. Mayhem like the 1883 "Bisbee Massacre," where five innocent people were killed during a holdup of the Goldwater and Castaneda store, was a thing of the past. There still remained, however, the explosive combination of a drifting male population that was largely under thirty years of age, and around the clock access to liquor, gambling, and prostitution. The "wild and woolly" atmosphere died hard, in part, in Lemuel Shattuck's St. Louis Beer Hall on August 19, 1903.

Night bartender Otto Schmidt had just locked the saloon's receipts in the safe, and L. O. Milless, the roulette dealer, was figuring his account, shortly before one o'clock that morning.

"I won and I lost." Milless matter-of-factly answered the question posed by faro dealer Tom Wilson, who had closed his game and was pacing in the middle of the floor. The only other men in the saloon were Louis Peluca and Joe Devitt, who lounged with their feet up on a table near the faro bank.

So calm was the atmosphere within the St. Louis Beer Hall that no one really paid much attention to the entrance of two men, wearing masks made from old hats in which eye-holes were cut. Otto Schmidt merely thought there had been a masquerade somewhere. But when the order came to "throw up your hands, all of you," the bartender "knew they were up against it," and he reached for the six-shooter behind the bar.

"No, you don't. Come out of there, you sonofabitch," ordered a short, heavy-set man, pushing his revolver into Schmidt's face.

As Schmidt stepped from behind the bar, the man took a shot at him. At the same time, the second man, a tall, thin individual wearing a torn red shirt and well-worn overalls, pointed his revolver at Milless.

"For God's sake, don't do that, man," the gambler exclaimed, raising his hands. Two shots blew the fifty-six-year-old dealer out of his chair. Instantly Tom Wilson drew his gun and began firing, frightening the men from the premises.

The first lawman on the scene, Town Marshal Charles Thomas, found Milless dead. He had been shot twice: one bullet passed through his right arm, just below the shoulder, penetrated his body and exited his back; a second shot barely scratched the old man's wrist. Otto Schmidt lay behind the bar, six-shooter in his hand, begging for a physician to attend his wounds. A bullet had grazed his right leg and passed through the calf of his left leg. Tom Wilson stood in the center of the room, gun in hand, "looking as cool as if nothing out of the ordinary had happened." Devitt and Peluca were cowering behind the faro bank.

The shooting was a vicious act, likened by the press[6] to the Bisbee Massacre—the two crimes would stand as Bisbee's worst. Shattuck's saloon, always orderly, was one of the most respectable establishments in town. In fact, during its early years Lemuel had acted as his own bouncer, ejecting troublemakers when the occasion arose. L. O. Milless was well known and liked, although not much was known of his past. It was said he had relatives in Ohio and may have been from Carthage, Missouri.[7] What was known about him is that he showed up in Albuquerque about 1873, gained a reputation as "an extra good billiard player," was robbed by Indians, contracted a light case of smallpox, and was involved in a shooting scrape with a Mexican, which caused him to head for southern Arizona.

Milless resided in Cochise County for twenty-one years, first being noticed among the sporting element in Tombstone. Upon decline of the silver camp, he moved to Bisbee and joined the staff of the St. Louis Beer Hall where he attained a reputation for "unflinching honesty behind the tables."

A reward of $500 for the arrest and conviction of Milless' murderer was posted by his lodge brothers, and Lemuel Shattuck matched the amount. On August 25 Shattuck, butcher E. A. Tovrea, and Joseph Martinelli, head gambler at the St. Louis Beer Hall, petitioned the City Council for the employment of night policemen.[8] Civic response was immediate. On September 1 Ordinance 31 was passed, allowing the hiring of as many night officers as the city deemed necessary to protect its citizens and businesses. The officers, whose salaries were set at $100 per month, would patrol from six p.m. until nine a.m.[9]

Murder was a common occurrence in mining towns, especially among ethnic groups and the sporting element. Bisbeeites barely blinked when someone was killed on Chihuahua Hill or in the tenderloin. But when that crime occurred in a respectable saloon, owned by a prominent businessman who was also a founding father, every resource was brought to bear to apprehend the culprits. City police, Cochise County sheriffs, and the elite Arizona Rangers took up the search. Within a month Cochise County lawmen had two suspects: Johnny James and Bert Noftz, both well known in Bisbee.[10]

James had worked at the Anheuser Saloon, and had been a stage hand at the Opera House.[11] Bert Noftz was an "attache" of the Opera House. The men were close friends and frequented the tenderloin. The sleuthing of City Marshal Harry Jennings revealed the motive behind the attempted robbery of the St. Louis Beer Hall.[12]

Apparently Noftz had been left in charge of the Opera House, while its manager, A. L. Manahan, vacationed. The day before the attempted robbery of the Shattuck saloon, Manahan returned to

The St. Louis Beer Hall was a long and narrow structure, built of rock and mortar. A small second floor existed over the cold storage room. A workingman's bar, the Shattuck and Keating saloon was modestly decorated with litho-

find saloon accounts in disarray. Hopping mad, the Irishman confronted Noftz, who admitted squandering money on Mabel Carlisle, an inmate of Cora Miles's bawdy house at 48 Brewery Gulch. Manahan probably would have accepted Noftz's promise of repayment had not matters gone from bad to worst. Noftz had been keeping the Carlisle woman at his quarters in the Opera House — in violation of city ordinance — and had boarded James and his woman. Manahan threw them all out, with the threat that if the money was not returned, he would come gunning for Noftz.

Down on their luck, without immediate income, James, Noftz, and Mabel repaired to the Wave Candy Parlor to lick their wounds. According to Charles Knapp, a seventeen-year-old employee of the establishment, the robbery of the St. Louis Beer Hall was hatched in a booth at the candy parlor.

At the time, Knapp didn't think much of the conversation he overheard. When news of the crime swept Bisbee, however, what he had observed and heard took on deadly meaning, and Knapp contacted Officer Jennings. The youth related the conversation as best he could, informing the police that the trio had definitely plotted the robbery, and both men appeared armed with pistols. The description Knapp gave of the men tallied with that of the robbers of Shattuck's saloon. One was short, heavy-set; the larger of the two was thin and wore a red shirt with pink stripes. Both men wore blue overalls.

James was immediately arrested and placed in irons at the city jail. Apprehension of Bert Noftz was not to be that easy. Following the attempted robbery, he and Mabel Carlisle had skipped to Cananea, Sonora. They found lodging at the American House, and Noftz took a job shingling the roof of the municipal building. He was working in that capacity when arrested and turned over to Mexican authorities for extradition by Deputy Sheriff George Chidester, Bisbee Marshal Harry Jennings, and George Asche in late September 1903.

When arrested Noftz broke down and cried like a child, bemoaning the day he first began to associate with James. He laid all the blame on his confederate.

"I knew this was coming when I saw you over here," Noftz told the lawmen. "I was afraid that sonofabitch James would give it all away."

With Noftz in custody of Mexican authorities, Harry Jennings and George Asche could not resort to strong-arm tactics employed north of the border to get a written statement from Noftz. Jennings resorted to the next best course — by implying if a statement was not signed, Bert would not be transported by train to the United States, but would be loaded in a buggy and driven to Tombstone at night. The lawman emphasized that it was a long trip and anything could happen along the way. Noftz's son, Hugh, could end up an orphan, cleaning cuspidors in Bisbee whorehouses. On October 12 Noftz penned his statement, which, curiously, was addressed to L. C. Shattuck:[13]

Dear Sir:

This is to certify that I, Bert Noftz, do hereby send you this letter in care of Harry Jennings to inform you that on the night of the attempted robbery that I was with John James. On that night I was an unwilling partner to the killing of Milless of which I was innocent. I did not know anyone was killed till two days later when I read it in the papers.

[Signed] Bert Noftz

While Bert Noftz awaited extradition, Bisbee authorities were attempting to sweat a confession from Johnny James. Arrested by Jennings on September 21, James was not charged with the murder of I. O. Milless until October 6, when Joseph Martinelli filed a formal complaint. Upon arrest, James was thrown into a dark cell in the city jail, his arms and legs shackled and fastened to chains affixed to rings in the concrete floor. The chains were so short that he could move his feet only a few inches. He received no food for four days and only filthy water to drink. During the next four days, city police provided "a few crumbs of dried and unwholesome bread," and tea so impregnated with soap as to nauseate and sicken the suspect. James suffered during the cold nights. He had no bedding except one old, worn, and very dirty quilt. Until his arraignment on

October 17, James was subjected to "vile and insulting remarks" from jailer John W. Mullen, who constantly told him that Noftz had confessed in Cananea, and implicated him in the murder of Milless.

While James broke down several times during interrogation by Jennings, and cursed Noftz, he steadfastly denied any knowledge or participation in the holdup of the St. Louis Beer Hall or the murder of Milless. In fact, Bisbee authorities never succeeded in wringing a confession from James. At his arraignment, James maintained his innocence and complained of his treatment at the hands of city police. On December 1 his attorney, John McGowan, filed suit for wrongful imprisonment and physical and mental abuse. Damages amounting to $5,500 were sought against City Marshal John W. Mullen, and his bondsmen L. C. Shattuck, Joseph W. Muheim, Theodore F. Metz, James Letson, and J. F. Lippert.

At the end of December Mexican authorities approved Noftz's extradition, and he was conducted to Tombstone by Cochise County Deputy Sheriff Dell Lewis. Long before arraignment, it was generally conceded that the treatment of the suspects by lawmen would be a major issue in the case. The Bisbee *Daily Review* had seen to that by stating on October 29 that Noftz's confession was obtained by threat, if not fabricated altogether by Harry Jennings and others. The paper added to its accusation by publishing a long list of "questions asked Noftz," with the suspect's answers—answers which denied any part in the attempted robbery and the murder of Milless.

As expected, when the preliminary hearing was held in Tombstone on the morning of January 7, 1904, the court promptly struck Noftz's statement as evidence on the ground that "it was improperly obtained." Because of publicity the case received and the personalities involved, defense attorneys argued that the defendants stood no chance of receiving a fair or impartial trial anywhere in Cochise County. The Court and district attorney agreed; a change of venue to Pima County was granted, and it was agreed that the defendants would be tried separately.

Bert Noftz's trial commenced at Tucson early on May 2, with

District Attorney Roscoe Dole of Pima County, and Bisbee attorneys George H. Neal, Benton Dick, and Fred Sutter representing the prosecution. John McGowan, W. M. Lovell, and Frank Doan appeared for the defense. Although the prosecution called thirty witnesses over the course of the two-day trial, the evidence presented was largely circumstantial. Otto Schmidt, Louis Peluca, and Tom Wilson failed to establish any connection between the killing and the accused. Defense attorneys called four witnesses and attempted to prove that prosecution witnesses had received money from L. C. Shattuck and "were to be paid a certain amount of monies if the conviction of the defendants was secured." Noftz's written statement, although inadmissible as evidence, was nevertheless damning to his opponent. The prosecution, attempting to counter its impact, argued in summation "that no innocent man would sign a document that would place him in the shadows of the gallows."[14]

At 4:00 p.m. on May 4, Noftz's case went to the jury. For nine hours it deliberated: some of the jurors felt the defendant was guilty of murder in the first degree, others stood for the lesser verdict of second degree murder, and a few favored acquittal. A compromise verdict was finally reached — Bert Noftz was found guilty of manslaughter, which carried a maximum penalty of ten years in jail. Judge Davis sentenced Noftz to seven years in the Territorial Prison at Yuma.[15]

The trial of Johnny James commenced on May 5, immediately after sentencing of Noftz. The trial, however, was of short duration. After the prosecution had offered its testimony, Defense Attorney Lovell moved for dismissal of the case on grounds that evidence failed to connect James with Noftz or the attempted robbery and subsequent murder. While Judge Davis ruled "that the evidence cast strong suspicion on James, it was not incompatible with his innocence, and no such connection as proven by the confession in the Noftz case [would be allowed]." Lovell's motion was granted and Johnny James was acquitted. Although the outcome of the trial surprised many, the *Daily Review*, a critic of the entire proceedings against Noftz and James, stated that "in the mind of many there has always been a doubt that these were the men who committed the act." Now James "will press his suit against Shattuck and his bondsmen for damages for cruel and inhuman treatment while confined in city jail," needled the paper.

Criticism by the *Daily Review* is strange in light of the fact that treatment of the defendants was typical of frontier law enforcement, where a crime had been perpetrated against property of a prominent citizen. Lemuel Shattuck, saloon keeper, merchant, and Bisbee founding father, had a loyal following. He had provided lumber for Bisbee homes, and built many. His beer was consumed throughout the Warren District, his resort a popular gathering place for men in their off-hours. Shattuck's knowledge and advice was sought on mining matters. As an owner of a large number of properties, in all probability a number of lawmen rented Shattuck cottages along Brewery Gulch. He had been instrumental in helping many Bisbeeites get started in business, acquire mining claims, or develop mineral deposits. In turn-of-the-century Bisbee — or in any other frontier locality — business was conducted with a handshake, loyalty was taken seriously. While fear of arousing Lemuel's displeasure might explain lawmen's zeal, it did not explain the criticism of the Noftz-James case by the *Daily Review*. The paper's criticism was in all probability rooted in the fact that Lemuel Shattuck was treading on the toes of Copper Queen management.

The drama of murder at Shattuck's saloon did not end with putting Noftz behind bars; the case would take a curious twist. Bert Noftz labored as a carpenter and upholsterer at the Territorial Prison, eventually becoming a model prisoner, which allowed him to practice his trade in Yuma. Never waivering in his claim of innocence, Noftz convinced the warden to petition Governor Joseph H. Kibbey to review the case. At the same time, Noftz's relatives in Douglas submitted compelling evidence to Lemuel Shattuck that the wrong man had been convicted of murder.

By May 1906 Shattuck was convinced that Bert had been railroaded by zealous lawmen. In his characteristic sense of fairness, Lemuel wrote Noftz "that he knew that he had nothing to do with it [the murder], and that he would do as much to get him out as he did to get him in."

Two years of a man's life had been wasted in prison. Noftz was impoverished, and once released, he would find it difficult to procure a job. To ease his transition, Shattuck requested that Bert "come to Bisbee the minute you get your pardon and any reparation that can be made, I will gladly offer."[16] Two years would elapse before anything was done regarding the Noftz case.

Governor Kibbey reviewed the case in October 1908, and concluded that Bert was innocent of murder, but "had knowledge of who committed the act." His sentence, which would expire February 14, 1909, was commuted to October 27, 1908, for good behavior. The following day, Bert Noftz stepped off the train at Bisbee, to be greeted by "several men of good standing," among them Lemuel Shattuck, "who tried their best to welcome him back and to assure him that they believed him innocent." Bert Noftz was joined by his family and lived out his life in Bisbee.[17]

Before the day of modern banking facilities, the Copper Queen Mercantile Company, founded in 1886 to satisfy a need for reasonably priced general merchandise, acted as a depository for funds and valuables. Deposits were received from miners, cattlemen, and other mining ventures, not only in Bisbee but as far away as Cananea, Sonora. The store also handled the once-a-month payroll of its parent company, the Copper Queen Consolidated Mining Company. By 1900 the "Merc," as it was called, was safeguarding much of the assets of the Warren District. Bookkeeping and the physical storage of funds became burdensome, which led to establishment of a full-fledged banking institution.

On February 10, 1900, William H. Brophy, manager of the Copper Queen store, M. J. Cunningham, an employee of the store since 1893. James S. Douglas, Jr., who had recently joined his father in the copper camp, and Ben Williams established the Bank of Bisbee, the first bank in Cochise County.

The Bank of Bisbee, with assets of $253,630, opened in humble quarters in the Angius Building on Main Street. Growth was spec-

Lemuel Shattuck poses between his two close friends: Baptiste Caretto to his right, Joseph Muheim on Shattuck's left.
(Caretto collection, Bisbee Mining and Historical Museum)

tacular and within two years the institution had acquired its own building, a two-story concrete and brick structure opposite the firehouse, where, according to the *Daily Review*, "the city council met, swore, told stories, chewed tobacco, and attended to business during intervals between stories."[18]

Because unwritten bank policy of the day held that "he who makes the loan, collects," the Bank of Bisbee was a conservative institution to say the least. It did not make loans where there was considerable risk — and the riskiest loans were those involving mining property. A fact that Lemuel Shattuck and every knowledgeable mining man knew.

But Shattuck had no choice. While he had accumulated a sizable fortune in real estate and had cash reserves, he did not have funds enough to straighten the crooked inclined shaft at the World's Fair claims, install an up-to-date power plant and hoist, and secure equipment and miners to develop the ore reserves that he, Grenfell, and Keating felt so positive about.

Shattuck literally had three strikes against him when he approached the Bank of Bisbee late in 1901 to secure a loan to develop the World's Fair claims. First, the claims were in an area beyond the limestone horizons where mines of the Copper Queen Company were extracting a fortune in copper. The claims were in a locality dissected by porphyry dikes, a place that many thought devoid of secondary enrichment. Brophy laughed at the prospects of finding carbonate lenses on the World's Fair claims of the size encountered on Copper Queen properties; the Bank of Bisbee denied the loan on grounds the World's Fair group could never produce enough copper to repay a sizable loan at twelve percent interest.

Shattuck was philosophical about being turned downed by the Bank of Bisbee. Intuitively he knew that vast amounts of copper lay on the property, and that someday he and his associates would profit from those ore bodies. What irritated him was the smugness of the Copper Queen crowd. They acted as if their mines were the only copper producers in the territory, and they had the monopoly on mining property at Bisbee. Money was always available to individuals in the Copper Queen hierarchy, but not to hardworking businessmen who also had interests in mining. Perhaps Bisbee needed another bank — a miners' and merchants' bank? He posed the question to Isabella, and she emphatically concurred.

Shattuck first presented the idea to Joseph M. Muheim, who had immigrated from Switzerland and settled in Bisbee in 1887. Like Lemuel, he had risen through the camp's commercial circle, from a lowly employee in uncle Frank Dubacher's brewery, to own a fine saloon, as well as real estate and mining claims. Muheim recognized a need for another bank, and embraced the idea. Equally enthusiastic was ex-Copper Queen mechanic Jacob Schmid, also from Switzerland. He had homesteaded land that was just beginning to develop into Bisbee's commercial district. And Shattuck's old friend, butcher and business partner, Lemuel J. Overlock, felt that the town needed another financial institution. None had knowledge of banking in the strictest sense of the word, however, but all pledged funds. They needed a professional banker to serve as cashier, or manager. Shattuck and his associates found that person in J. T. Hood, whom they lured away from the Bank of Bisbee. Although a young man, Hood had ten years' experience in banking, including two with the Copper Queen institution.

Quarters were rented at the rear of Henkel's jewelry store in the new Letson Building on Main Street. A Mosler patented screw door safe was obtained. Constructed of the finest steel, the one-ton safe was conceded to be burglar proof. Its round door had a three-movement time lock of Yale and Towne manufacture, and contained no bolts such as used in square safes. Two heavy, steel screws locked the door to corresponding threads in the safe so tightly that it was impossible to insert liquid explosives or wedges. A concrete and brick vault, with walls twenty-two inches thick and an iron roof, was built for protection against fire. Its door was also of Mosler design.

The Miners and Merchants Bank published its articles of incorporation on June 18, 1902.[19] The bank's opening, scheduled for mid-July, was delayed by construction of the vault.[20] When the institution's quarters were finally in order, its doors were thrown open to the public at 9:00 a.m. Wednesday morning, July 30, 1902. Fully capitalized at $50,000, the Bank had thirty stockholders, all prominent men in Bisbee and Cochise County business circles. Undoubtedly a sad day for the Bank of Bisbee, the Copper Queen–backed *Daily Review* gave no hint of irritation when it announced the officers of the new financial institution: L. C. Shattuck, as president "is one of the best known men in this part of the territory and his name stands for the highest degree of business integrity, financial responsibility, and honorable dealings. No deserving person has ever been denied a favor at Lem Shattuck's hand, and he has helped many to the road of prosperity." Vice-president Muheim "likewise . . . is a man of means and his word is as good as his hand. He has many friends and no enemies, and his name for square dealing is without a blemish." And Cashier J. T. Hood "has proven his fitness for the position here and elsewhere." Directors Jacob Schmid and L. J. Overlock "are well known as men of responsibility, and are both possessed of considerable means, which they accumulated by hard work, shrewd business methods and fair and generous treatment to all." It would be hard to find and associate together finer men, who stand better in the community, than the board of directors of the new bank. We predict a prosperous future for the Miners and Merchants Bank."[21]

Despite the glowing introduction, everything about the Miners and Merchants Bank was modest — its quarters, its capitalization, and even the way it advertised. "A share of your patronage solicited," read the Bank's advertisements in the *Daily Review*. Its president, cashier, and directors were all men possessed of acumen. It would be unfair to say that Lemuel Shattuck had no experience in banking. He certainly knew accounting. Ever since opening the St. Louis Beer Hall, he had cashed miners' pay vouchers, issued tokens for drinks, and extended credit to preferred patrons. Saloons, restaurants, boardinghouses, and whorehouses bought Anheuser Busch beer from Shattuck, most on open account. Like every saloon in the Far West, the St. Louis Beer Hall also maintained an accounting system for its gambling concession, whereby roulette and faro bank dealers received a fair share of the proceeds of the games.

Starting out as the third bank in Cochise County (behind the Bank of Bisbee and its sister institution, the Bank of Douglas), the Miners and Merchants Bank grew at a phenomenal rate. In less than a year it had deposits of $300,000, loans and discounts totaling $150,000, and cash reserves of $200,000.[22]

Within two years the Miners and Merchants Bank had outgrown its cramped quarters in the Letson Building. In March 1904 the Bank purchased the site of the Anheuser Saloon from Otto Geisenhofer for $23,000. Shattuck purchased the saloon's lease from its owners, Boston and Brown, tore down the building, and in late August began construction of a new two-story, pressed brick building to house the Bank.[23] The splendid new quarters were but one door away from the Bank of Bisbee, which had denied Shattuck a loan three years previous.

Success of the Miners and Merchants Bank was in large measure due to the personality and guiding hand of its president. Years of selling lumber and liquor honed Shattuck's judgment of men. Uncanny in his ability to detect both failings and virtues in men, he left his office door open to all — business executives, capitalists, mining experts, old prospectors, gamblers, and cattlemen. He always had time to listen to the woes of old sourdoughs, miners and others less fortunate. He measured worth not by appearance, but by the qualities of the Old West — honesty, fair play and responsibility. He was apt to give more time to the little fellow than to the man-of-means.

Many times he loaned large sums on a handshake. Once when he did not ask a borrower to sign a note, the man asked why. Lemuel granted the $10,000, stating that if he misjudged the man's character and he did not repay, his word would be no good and a signed contract would be just as worthless. Lemuel loaned thousands of dollars to long-time acquaintances in Arizona and Sonora, but never passed over the small borrower. He provided money to tide farmers over difficult times, start small businesses, build corrals and fences. At meetings of the Bank's board of directors, Shattuck always insisted that unpaid loans be referred to him. More often than not, he would take out his checkbook and write checks to cover the small outstanding balances. For the most part, these borrowers made an effort to repay. In those days a man's word usually was good.

The mining camp of El Tigre, Sonora, Mexico, about 1908.

FOUR

BUCCANEERING FOR EL TIGRE

BEFORE THE Miners and Merchants Bank was founded, Shattuck was known throughout southeastern Arizona as "Lem": saloon owner, merchant, owner of promising mining properties. The Bank changed this, consolidating his influence and increasing his prestige. As rival to the Bank of Bisbee, the Miners and Merchants Bank rapidly became a depository for businesses not connected with the "Copper Queen Crowd." As its name signified, the Bank catered to the interests of Bisbee merchants and independent mine owners, the latter just beginning to emerge in the Warren Mining District. In heading the institution Shattuck attained a level of power backed by financial ties and real estate holdings. The institution provided him a credible base for movement through the world of finance. Affiliation with the Arizona Banking Association and establishment of ties with larger banks, such as Boston's National Shawmut and the American National Bank of Kansas City, brought introduction to industrialists seeking western mining properties.

The Bank, headed as it was by a man experienced in mining, was from the outset a magnet for prospectors and mining promoters seeking financial aid to develop claims, or introductions to men of means who might bond or lease their prospects. Shattuck was besieged by men claiming to have struck it rich. Over the course of twenty years, he had seen nearly every type of mineral occurrence, heard the trumpeting of prospectors, and dug enough ore from his own claims to separate reality from fantasy. He could determine the worth of most properties. Shattuck examined ore specimens, listened in a reserved manner, and courteously rejected most of these visitors. Occasionally a miner appeared with rich minerals or assay reports that no professional mining man could overlook.

Such was the case shortly after the opening of the Miners and Merchants Bank in 1902, when Benjamin F. Graham, a small, rotund mine promoter who also ran the Bisbee-Naco Stage Line,[1] strode into Lemuel Shattuck's office and placed two canvas bags on the banker's desk.

"What do you have there, Ben?" Shattuck asked, motioning the man to sit down.

"The richest ore you ever saw," replied Graham, pulling up a chair.

Graham untied a bag, reached in and drew out several pieces of dark, ferruginous quartz and handed them to Shattuck. Lemuel examined the rock, drew a pocketknife from his desk, and scraped away some of the powdery iron oxide. The ochre-colored specimens were laced with arborescent gold, and from their weight, he judged the specimens to be 50 percent metal. Graham grinned, and pulled from the other bag pieces of hard "greasy" quartz, having little if any oxidized iron.[2] These, too, were speckled with gold. They were specimens unlike any Shattuck had seen.

"Someone must have hit a pocket?" quizzed Shattuck cautiously.

"A pocket, hell yes. I have tons of ore like this, from two veins that run for thousands of feet, and they're full of pockets. The entire outcrop shows high values in gold and silver as well as secondary minerals."

"Where are these veins?" the banker asked, forcing back a welling of excitement.

"El Tigre Suertudo," Graham responded, "sixty miles south of the border."

"Lucky Tiger," Shattuck translated. "Judging from this ore,

someone's lucky and its not a cat," he mused. "Tell me about the find."

What Graham told Shattuck was typical of discoveries of Western mineral deposits, in that the strike was made by accident. El Tigre was not named after a cat. "Tiger" was a dog whose persistent appetite and unflagging curiosity led him into strange adventures, the climax of which was reached on May 24, 1901, when he encountered a mountain lion in Chino Canyon, in the heart of the Teras Mountains of Sonora, Mexico. His master was James H. Taylor, rumored to be an outlaw on the run from Texas, who had devoted his summer prospecting the headwaters of the Yaqui River with friends, Edward M. Sturges and H. Stine.[3]

While Taylor examined the dry, precipitous, rock-strewn slopes of the canyon, his dog nosed along the stream bed some distance below. Suddenly the dog came upon a mountain lion. As hostilities were about to commence, Taylor grabbed up a rock and threw it at the cat. He secured another rock and was about to throw it when the animal, confronted with two enemies, retreated. Taylor was left with a rock in his hand, a rock unusually heavy for its size.

Taylor broke it open and found it literally plastered with gold. He searched about and found similar rocks: definite sign of a wash of ore from a higher source called a float. Taylor picked up random samples of gold ore from the talus until he encountered, high on the canyon wall, a quartz vein containing gold and native silver that ran for hundreds of feet. Taylor and his dog Tiger had stumbled onto one of the richest gold deposits yet chronicled in Sonoran mining history.

Gathering up the best "show rock," Taylor hastened to find his friends. In camp the men gloated over the rocks, hardly believing what they had found. They knew, however, they would need verification of ore values if they were going to find financial backing to develop a mine, and they hastened to Bisbee to seek out assayor George C. Clark, who analyzed the ore. The assay revealed a fabulously rich strike, which caused the discoverers great anxiety.

The men hesitated to file for Mexican mineral rights for fear that in some way word of the richness of the find would leak out and they would be beaten out of their title by a stampede of prospectors. Miraculously their discovery was kept secret until they secured the "denouncement" (Mexican mineral rights) to their property.

Taylor, Sturges, and Stine were without means, unable to finance even meager exploratory work, and they sought out six men who were operating a stamp mill on the Bavispe River: A. C. Riordan; Alex Grant; James Allen; and the Suits brothers, Charles, Edward, and James. They entered the venture as partners and located the Combinacion claims, consisting of 148 acres just north of El Tigre Suertudo claims.[4] Together, the men had funds enough to drift into the veins at several points, and extract enough ore to run tests at their mill. Those tests determined that gold and silver values persisted in depth and were uniform over the extent of the outcrops.[5]

El Tigre was rich, but it had one drawback: it was forty-five miles south of the border, and thirty miles east of Ysabel station on the Nacozari railroad, virtually on the summit of the Teras Mountains. Rough, dissected terrain separated the gold deposit from the nearest railroad. Considerable capital would be needed to blaze a wagon road into the property and develop the mine. That is when the men sought out Bisbee mining promoter B. F. Graham. Beautifully streaked with gold and native silver, El Tigre ore portended great wealth. The mineral specimens, some $6,000 worth brought into Bisbee, captivated Graham, and he beat a path straight to the Miners and Merchants Bank and sought out Lemuel Shattuck.

While Shattuck was not a professionally trained geologist — his knowledge of mineralogy had been gained in the field — he nevertheless instinctively knew the chemistry of mineral deposits: that surface occurrences could assay exceedingly rich, but lean with depth. He had to agree that Graham's specimens were the richest he had seen, but beyond that he declined to voice an opinion until he examined the property, viewed the context of mineralogy, and weighed the logistics of supplying the mine and shipping its ore or concentrate to the nearest smelter.

Graham proposed an immediate trip to El Tigre, which Lemuel accepted, and they spent a week inspecting the properties and conferring with Taylor and his associates, who were hard at work prospecting the mineral outcrops. What Shattuck observed proved Graham's assessment of the property. Two well-defined fissure veins ran for several miles, although in places they were covered with debris owing to the precipitous terrain. Rich ore containing visible gold and silver outcropped at the surface. The principal ore, however, appeared to be a "steel" galena, carrying silver sulphides associated with antimony, zinc blende, and wulfenite. A fine-grained pyrite, high in gold, was disseminated throughout the vein and surrounding rock. El Tigre and Combinacion claims were indeed rich, and Shattuck consented to help Graham promote the claims into a profitable mine, or dispose of the properties to a middleman. Either way the claims would generate a handsome profit that would make wealthy men of the discoverers and pay a sizable commission to Shattuck and Graham. The Miners and Merchants Bank would handle the escrow, at one percent, if the property was sold. Should the owners desire to form a corporation, the Bank would handle the process of incorporation, as well as the sale of stock. With this agreement, formalized with a handshake, Lemuel Shattuck took his first step into the world of high finance and mine development.

As soon as they returned to Bisbee, Graham and Shattuck went to work. In February 1903 Graham obtained a one-year option on El Tigre and the Combinacion properties, and organized a Mexican corporation, The Tigre Mining Company, S.A., to acquire title to the properties. At the same time, an American company called the Lucky Tiger–Combination Gold Mining Company was incorporated under the laws of the Territory of Arizona.[6] The venture would be capitalized with 65,000 shares of stock at par value of $10, one-half the price payable in advance, and one-half subject to call of the directors. This money would be used to buy out the original owners. An advertising program was immediately launched through the *Daily Review*: "Procrastination is the thief of time," read one advertisement. "If you don't invest at once in Lucky Tiger gold stock, you will discover in a very few days that this same fellow is also the thief of your money. Invest at once and get in on the ground floor."[7] The ad pulled in investors.

While considerable stock was sold in Bisbee and Douglas — Shattuck, Graham, and Lee Benton each invested $50,000 — the promoters knew that the best market for the Lucky Tiger was in the Midwest and the East, where capitalists thirsted for western mining ventures. In hopes of tapping that market, Shattuck provided Graham with letters of introduction. One such letter, directed to the president of the American National Bank of Kansas City, painted Graham as "a gentleman of the highest integrity, soundest business judgment and thoroughly responsible in every way." Most important, Shattuck's letter stated that the mine Graham represented "was considered by everyone here [in southeastern Arizona] the biggest thing in gold mining ever found in these parts, and one of the biggest in the world."[8]

It was a heady recommendation that naturally gained the confidence of a number of prominent attorneys, bankers, and farmers of Kansas City and St. Louis, who clamored to see the property. A railroad car was chartered and twenty Missourians transported to Bisbee in early March 1903 to view the investment. Professional mining men, able to render a reliable opinion regarding the property, were recruited. Colorado mining experts Gus Landeau, G. W. Dollis, James Thompson, and A. W. Bateman arrived in Bisbee and were booked into the Copper Queen Hotel. Bisbee assayer and engineer, George C. Clark, would also accompany the group to prepare a prospectus on the property. Bisbee *Daily Review* mining columnist Frank Aley stood ready to chronicle the expedition. Guided by the "Twin Princes of Hospitality," Lemuel Shattuck and Lee Benton, a mining man who had a hacienda on the Bavispe River close to El Tigre, the party left Douglas the first week of March 1903. For six days they poked about the claims, hefted and cracked rock, swapped tales, smoked Mexican cigars, drank, ate well, and talked of the deposit with the discoverers.[9]

The investors came away thoroughly impressed with the potential of the property. And the experts endorsed the investment. "I consider El Tigre a very good prospect, and with proper development and management, it ought to be a paying proposition," stated G. W. Dollis, a miner and assayer from Florence, Colorado. James Thompson, a practical mining man from Breckenridge, Colorado, concurred. "It is a proposition that will require an outlay in order to properly develop it. It is in a wonderful mineral belt, some six or seven miles in length. I believe that with about one year's development, wherein the necessary means are at hand, it will become a large mine and a dividender, that will justify the investment of $650,000." George C. Clark also endorsed the potential of the property.[10]

The time came for Shattuck to voice a public opinion regarding El Tigre, and he did so through the mining column of the *Daily Review*. "It is the best proposition I ever saw, and I have seen nearly all of them," Frank Aley quoted Shattuck as saying. Aley added the observation that Lem Shattuck "has perhaps the soundest practical idea of mineralogy of any man in Cochise County."[11]

The midwesterners were also convinced, and on June 15, 1903, they invested $600,000, purchasing all the treasury stock of the Lucky Tiger–Combination Gold Mining Company, as well as whatever blocks of common stock they could wheedle from Bisbeeites. In short they bought the mine. The deed of sale specified a down payment to the original discoverers of El Tigre and Combinacion claims of $120,000 on June 15, 1903, the balance of $480,000 in installments as follows: $120,000 on or before October 16, 1903; $120,000 on or before January 16, 1904; the remaining $316,800 principal and interest extended over eighteen months. All payments would be in U.S. gold coin and made through the Miners and Merchants Bank in Bisbee. To protect stockholders against loss in event of default, the contract also specified that in case of failure on the part of the company to meet payments, the original owners had the right to form another company and protect stockholders of the Lucky Tiger Mining Company by issuing stock in the new company.

The Lucky Tiger–Combination Gold Mining Company announced its officers: B. F. Graham, president and general man-

Lemuel Shattuck standing on an outcrop of gold-silver ore at El Tigre. This photo, one of Lemuel's favorite, was taken by George C. Clark in early March 1903. (Fathauer collection photo)

ager, at a salary of $800 a month plus expenses; J. K. Davidson, vice president; William Hargrove, secretary; A. L. Harroun, treasurer. These men were connected with the American National Bank. J. T. Hood, cashier of the Miners and Merchants Bank, was assistant treasurer. Board members from Kansas City and St. Louis were Harroun, William J. Morse, Davidson, W. A. Moses, H. Vanderslice. The Arizona board members were Lee Benton, Graham, Shattuck, Hood, and J. E. Suits.[12]

With exception of Hood, the Arizona board members were knowledgeable mining men, and were concerned that the midwestern majority stockholders, who lacked mining experience, might not listen to the westerners in determining company policy. The "Kansas City Crowd," as they would soon be called, denied that would happen, and proposed the Arizonans join them in what was termed a "voting trust" that would serve to protect the interest of the latter. And so it was agreed at the first board meeting and most members departed happy. Lemuel Shattuck, however, had misgivings. He was a realist who knew that once the mine was developed and the indebtedness retired, the midwesterners would undoubtedly forget their promises. They owned the majority of stock and would call the tune to which everyone danced. For the moment, however, everything was amicable.

On March 21 the *Daily Review* congratulated B. F. Graham and Company "on the successful conclusion of their negotiations for the disposal of the Lucky Tiger. It was a deal of exceptional magnitude, and as such, is always a difficult piece of manipulation, considering the limited time in which the promoters usually have to work. Along with B. F. Graham, Major Hargrove, Lee Benton, Lem Shattuck and other gentlemen of experience and influence contributed materially to the successful outcome of the negotiations."[13]

Sale of stock enabled the Lucky Tiger Mining Company to commence full scale exploration of its quartz veins and to link its mines with the railroad. Throughout spring 1904 Graham was everywhere, supervising and planning. He augmented his force of 100 Mexican miners by hiring 200 Yaqui Indians, who were put to work blazing a wagon road, at a cost of $35,000, to Ysabel station, the shipping terminus on the Nacozari railroad.[14] When not at the mine, he was in Bisbee conferring with mining equipment manufacturers. In March Graham procured a concentrator to process low-grade ore. As the foundations of the mill were being poured, mining went on ten hours a day, seven days a week. For months every train leaving Douglas carried machinery consigned to El Tigre. Strikes of rich ore were daily occurrences. In the meantime, only "first-class," free-milling ore, running $8,500 to $10,000 per carload, was shipped to the El Paso smelter. Value of this hand-sorted and cobbed ore was more than sufficient to pay all operating and development costs. In 1904 the mine produced $236,033 in revenue with a net profit of $92,482. Lucky Tiger stock soared to $30 a share. Investors clamored for the stock, but most stockholders hung on to their shares, some stating that they would not part with their stock for a hundred dollars a share.[15]

The mill was finished in late January 1905, and the first trial runs were made on February 2. As far as reduction works go, it was crude: steam-powered, a wood-burning boiler, and the water obtained from mine runoff, which at times was insufficient. Milling capacity was 115 tons per day, which at 17:1 rate of reduction produced about seven tons of concentrates a day. Tailings carrying about fifteen ounces of silver per ton were piled in a nearby gulch until such time as the company supplemented the mill with a cyanide plant.[16]

Production at the mine exceeded expectations. The Lucky Tiger–Combination Mining Company was able to meet—and exceed—its payment schedule. By June 1905 the balance of the note held by the original owners stood at $288,000, and another payment of $48,864 was due on July 6.[17] Just as Shattuck surmised, with reduced indebtedness came a change in attitude on the part of the "Kansas City Crowd."

At the company's board meeting on April 15, 1905, Graham was reduced to vice president, a position that had practically no power. Kansas City banker E. C. Sooy was elected president. Shattuck, Graham, and Benton were voted off the board of directors, their seats taken by Missourians, and the company headquarters, books and records were moved from Douglas, Arizona, to Kansas City, where an executive committee of men with no mining knowledge was formed to manage the company. This executive committee, headed by attorney O. V. Dodge, had no intention of running the mines. It was rumored in financial circles that the Kansas City men planned to "knock" the value of stock, and buy all shares not in their control. They would then sell the mine, negotiation for which had been conducted in secret — with whom was not precisely known. Some say John D. Rockefeller was interested in the property; others believed it was Phelps-Dodge Company. The price was rumored to be $1,500,000, and the executive committee planned to pay themselves $300,000 commission upon final sale.[18]

Graham, Shattuck, Benton (who owned $150,000 worth of Lucky Tiger stock), and other Arizona stockholders were furious, but virtually powerless to stop the Kansas City crowd. Lemuel Shattuck felt the midwesterners were not only slippery, but did not know the true worth of the Lucky Tiger Company, bent solely on reaping a quick reward. The Kansas City Crowd nevertheless controlled the company, and could ride roughshod over western stockholders, regardless of any "voting trust."

B. F. Graham was not as fatalistic as Lemuel Shattuck. For years he had devoted all his energies to promoting El Tigre; he had even vacationed with his family at the mine. Without a doubt, El Tigre was his first love, perhaps an obsession, and Graham would not be bilked out of a mine worth millions by a group of scheming Missourians. He vowed to make the July 6th payment as difficult as possible, in the hope the company would default.

Graham germinated one of the most astounding plots ever hatched since the days of the freebooters. Early in May he tendered his resignation as vice president of the Lucky Tiger Company, contacted the original owners of the mine and convinced them the Kansas City Crowd was going to drive down the value of Lucky Tiger stock, buy it up, and sell the mine. Taylor, the Suits brothers, A. C. Riordan, H. H. Stine, and Alex Grant vowed to support Graham's attempt to counter the Missourians. Graham then proceeded to Cananea, Sonora, and on June 27, 1905, organized a Mexican corporation known as the Ensenada Mining Company, whose stockholders were the original owners of El Tigre and Combinacion mines.[19] Graham proportioned each man's share of stock, purportedly based on his share of the balance remaining on the contract (approximately $288,000). But the new percentage rate did not tally with that specified in the original contract of sale. Furthermore, Graham failed to notify both the Lucky Tiger Mining Company and the Miners and Merchants Bank of the reapportionment, and planned to require that the July 6 payment be made in gold, according to contract, a point that had not been adhered to for previous payments. It was a maneuver designed to make the payment of $48,864 "defective." Once the default had occurred, it was Graham's plan not to resort to court action to right what he conceived to be a wrong, but to seize the mine under the terms of the contract.[20]

Previous to Graham's resignation as vice president of the Lucky Tiger–Combination Gold Mining Company, Lemuel Shattuck had informed the company that he and other Arizona stockholders had more than $8,000 on hand at the Miners and Merchants Bank which they would contribute toward the final payment due July 6, 1905.[21] Therefore, on July 3 the company directed the American National Bank to wire the Bisbee bank $40,000.

That same day President E. C. Sooy and his attorney, William Morse, arrived in Bisbee to see that payment was properly made. They called at the Miners and Merchants Bank and asked for Lemuel Shattuck. Cashier J. T. Hood informed them that Shattuck was vacationing with his family in Erie, Pennsylvania, and volunteered to help them. Sooy requested that a credit be given for the money wired from Kansas City, as well as the $8,000 that Shattuck stated was on deposit at the bank. To his surprise, Hood could not account for the $8,000 from Arizona stockholders. Sooy then requested drafts for the $40,000 so he might deposit the money in another bank. Hood replied that he could issue the amount only in currency, which was accepted, placed in a valise obtained from the Copper Queen Hotel, and transported to the Bank of Bisbee.

The concentrator of the Lucky Tiger–Combination Gold Mining Company about 1908.

Cashier Hood's refusal to accommodate them raised the suspicions of Sooy and Morse that all was not right in Bisbee, and they wired Kansas City for an additional $8,000, to be sent to the Miners and Merchants Bank. Fearing a "hitch" in the receipt of this money, Sooy requested that the American National Bank wire another $8,000. Within a day Sooy and Morse had $8,000 in excess of the amount due July 6.[22]

In the meantime, gold coin reserves at Bisbee banks began to dry up. Someone was at work exchanging large amounts of currency. The Bank of Bisbee informed Sooy that it would be impossible to exchange his currency and credits for $48,864 in U.S. gold coin. He directed the bank to wire the San Francisco Mint for an equivalent amount in gold, which arrived in Bisbee via Wells Fargo on the morning of July 5, and was carried under armed guard to the Miners and Merchants Bank.[23]

The $48,864 in gold was presented to the cashier, along with a memorandum specifying the amount due each of the original owners, proportioned "exactly as former payments . . . and in accordance with the interests of each in the original bond." The note stated that the money was subject to the order of the original owners. Hood refused to accept the money until he consulted J. M. O'Connell, the bank's attorney. Following the conference, Hood took the money and presented Sooy a receipt.[24]

B. F. Graham arrived at the Miners and Merchants Bank at two o'clock on the afternoon of July 6, announced that "he was attorney in fact" and authorized representative of vendors of the Lucky Tiger–Combination Mining Company, and demanded the payment, not in sums specified by the original contract, but according to his apportionment. Cashier J. T. Hood stated that $48,864 in gold was on deposit in the bank to the order of persons represented by Graham, but only in proportions specified in the contract of 1903. Attorney William Morse, who was present, thereupon demanded that Graham execute and deliver a proper receipt. Graham refused to do this and stormed out of the bank to put in motion the second part of his plan, the seizure of the mine.

Graham had hired John J. Brooks, a lieutenant of the Arizona

Rangers, to raise a party of men to back the seizure of the Lucky Tiger Mine. Resigning his commission, Brooks purchased from the Copper Queen Store in Douglas revolvers, rifles, and ammunition enough to equip a sizable military company, which he stashed at the Roy Hotel. With the help of Red Seely, an ex-lawman, Brooks recruited a force of ten gunmen.[25]

These men accompanied Graham to Bisbee on the morning of July 6, and waited outside the Miners and Merchants Bank while their employer haggled with Sooy, Morse, and Hood. In fact, Graham and his gunmen, according to the *Daily Review* of July 8, "stood around outside the bank until midnight like a lot of mysterious school boys."[26] Returning to Douglas, Graham, his lawyer D. A. Richardson, Brooks, and the men picked up the weapons stored at the Roy Hotel, and left town the morning of July 7 aboard a chartered train bound for Ysabel station (arrangements having previously been made with the customs agent to pass the arms into Mexico). At Ysabel station the party boarded waiting coaches and arrived at El Tigre late in the afternoon.

According to Charles Allen, a mine employee, Graham's party arrived "armed to the teeth," each man carrying a sixshooter and a number of rifles. They went straight to the mine office. Graham informed Superintendent T. J. Wiley "that the Kansas City people were down and out," they had defaulted in making payment on the property, and it had become the property of the Ensenada Mining Company, of which Graham was president and general manager. Wiley made not the slightest protest and placed the mine in Graham's hands. The party stacked their arms in the office, and guards were placed at different points along the approaches to the mine. Thus with only the flimsiest technicality serving as an excuse to seize the mine, the Ensenada Company, under Graham's management, forcibly took possession of a mine worth possibly as many millions of dollars as there were buccaneers in the raiding party.[27]

The Kansas City stockholders were as tough as Graham — and they were not strangers to Mexico. Some were heavy investors in the Stillwell railroad enterprise as well as in other mining projects south of the border.[28] They were not at all daunted in seeking justice in Mexican courts. As President Sooy stated to the Douglas *Daily International-American*, "They were as ready to try their rights in Mexico as in the United States and felt confident that final results will be the restoration of El Tigre to its rightful owners with full reparation for all the expense . . . incurred in defending the property." To represent their side of the controversy, the Kansas City people hired D. J. Haff, a pugnacious Mexico City lawyer who had built a reputation for representing Anglo-American interests in Mexican courts.[29] There followed one of the hardest fought legal battles for possession of mining property that Mexico had ever witnessed—a battle characterized by chicanery, bribery of judges, accusations of filibustering, and brilliant legal maneuvering.

Attorney Haff opened the battle in August 1905 by filing a criminal action before Judge Enrique O'Farrell, of Moctezuma, Sonora, charging that Graham had invaded Mexico with an armed posse for the purpose of taking forceful possession of El Tigre and Combinacion mines.[30] Armed intrusion with intent to seize property was indeed a serious charge, no less than filibustering—which under Mexican law carried the death penalty. It was also a charge fraught with political and economic implications. Here was an incident that could be used by the Mexican government, if it so desired, to restrict foreign investment and travel in the country. Equally critical was the incident's effect on mining in Mexico. Graham's actions had interfered with closure of a number of sales of mining properties, among them the sale of the York Mines, near El Tigre. American investors were reluctant to place their money in Sonoran mines, and bitter were the feelings of mine owners toward Graham.

Judge O'Farrell went to El Tigre with a police escort and convened a two-day court of inquiry. Graham and his men testified that there had been no armed intrusion, nor seizure of a mine. It was true, they acknowledged, they had arms, but no more than ordinarily carried in isolated parts of Mexico. That testimony probably would have been accepted, if one of Graham's men, who had been drinking, had not given the whole scheme away. Judge O'Farrell seized the mine, arrested Graham, Brooks, and Wiley, and restricted movement of El Tigre ore.[31]

R. S. Vickers, prominent banker and mine owner of Moctezuma, signed a bond securing release of Graham, Brooks, and Wiley, and attorney Richardson bribed the district judge at Nogales to dismiss charges against Graham and his associates and overturn O'Farrell's decision. A few hundred dollars did the trick, and the Ensenada Company was appointed receiver for the Tigre property pending a hearing of the case by the supreme court of Mexico City. Late in August Richardson and Graham again returned to Judge O'Farrell's court to file suit seeking annulment of the contract of sale between the Lucky Tiger–Combination Gold Mining Company and the original vendors of the mines, on grounds that the purchasers failed to make either payment or tender of payment of amount due under contract of sale.[32]

It is likely that Graham and his attorneys bought O'Farrell, for the judge who had responded speedily before now moved at a snail's pace. Furthermore, no oral testimony was permitted in court, all facts being gathered from interviews with witnesses. Sixty days were spent laboriously transcribing the data for Judge O'Farrell's perusal, during which time the Ensenada Company mined hundreds of tons of rich ore. Had the buccaneers been able to market this ore and obtain money to continue the bribery of high officials, the outcome of the case might have been different. But officials of Phelps-Dodge Company, who operated the Nacozari Railroad, refused to move the ore a foot from Ysabel station. On top of this, the mill at the mine was inoperable due to a broken main shaft, and the camp was nearly out of provisions. Merchants at Oaxaca, where the company bought supplies, refused to honor orders.

Graham had scored the first blood in the battle for El Tigre, but in so doing had alienated much of his support. On the night of September 19 every person having an interest in the Ensenada Mining Company met in Douglas. The meeting, described as "stormy," was held to reconcile some members of the company who were not satisfied with Graham's conduct. A. C. Riordan, James A. Taylor, Alex Grant, and C. H. Suits, who represented slightly less than half the capital stock of the Ensenada Company, branded Graham's actions in taking the mine by force as "void, unauthorized and illegal." Although not a member of the Ensenada Company in the sense of being one of the original discoverers, Lemuel Shattuck owned considerable Lucky Tiger stock. His interest in the property brought him to Douglas within a day of returning from his vacation in Pennsylvania. Present at the meeting, Shattuck was appalled at seizure of the mine on any pretext. The banker let Graham know that his actions were those of a madman, jeopardizing mining interests in Sonora, and creating a situation that could endanger the credit and responsibility of a public banking institution. Shattuck expressed a determination to "get out of the muddle."[33]

Even William C. Greene, president of the Greene Consolidated Copper Company, suspected of backing Graham in hope of later buying El Tigre, denied any interest in the property. "I do not approve in any way of Mr. Graham's course," he told the Douglas *Daily International-American.* "On the contrary, I consider the seizure of the mine a mistake on Mr. Graham's part, and in my opinion there can be but one outcome to the litigation."[34]

Greene was right. By the first week of November, Attorney Haff and his Mexican lawyers had succeeded in transferring the case to the superior court at Hermosillo. Judge O'Farrell was called before the court to show cause for the delay in trying that part of the case under his jurisdiction. He was reprimanded and the case withdrawn entirely from his hands. On December 29, 1905, Esmael Elizondo, Judge of the First Instance of the District of Moctezuma, ruled that the payment of July 6 was legally made when Mr. Sooy deposited the money into the Miners and Merchants Bank to the credit of the original owners. Going to El Tigre with a police escort, Judge Elizondo turned the mine and all its property over to B. A. Suitz, as representative of the Lucky Tiger Company, until the federal court rendered a final decision. That decision was handed down in June 1908, returning the mine to the Missourians and awarding them $275,000 damages against B. F. Graham. The Lucky Tiger Company filed a lien against his property, and his real estate in Bisbee and Douglas was seized by the sheriff and sold at auction to satisfy a small portion of the judgment. The greater amount of the damages, however, was never collected, for Graham fled to British Columbia, fearing criminal charges from the Mexican government.[35]

The buccaneering of El Tigre was widely reported by the press, the drama of the case obscuring the fact that in the span of a few years the mine earned a reputation for being the richest newly-developed property in the Republic of Mexico. Shattuck may have been frustrated at Graham's actions, but he knew that time would extract him from the "muddle," and El Tigre would again run profitably. As a financier who had purchased a $50,000 block of stock, he did not relinquish equity in a mining company that still had years of life, but remained a shareholder for several decades. He was receiving dividends from the company as late as February 1929.[36]

Although El Tigre had its distressful moments for Lemuel Shattuck, in later years he spoke fondly of the property, drawing often from the family photo album a picture taken of him during the February 1903 trip, standing against an outcrop of ore at the property. His pride was justifiable: El Tigre was his first successful venture into mine promotion, and the mine had a long and distinguished history. It not only withstood a wholesale plundering of some of its richest ore, but paid its purchase price, its development and litigation costs, and emerged from the storm a splendidly developed property. Once back in control of its Kansas City owners, El Tigre paid its stockholders no less than two percent per month interest on their investment, and in some months five percent.[37]

The mine, however, had its ups and downs. On March 2, 1911, El Tigre was again seized, this time by a band of *Maderistas* under General José Blanco, who stripped the company store, and ran off with several hundred rifles and 50,000 rounds of ammunition. The rampaging of Pancho Villa in 1915 forced the evacuation of El

Tigre, but the mine was not molested.[38] Eighteen months later it was taken over by the Sonoran state government, which turned it back to its American owners.[39] Regardless of the tide of revolution that swept Sonora between 1911 and 1929, El Tigre produced steadily. Between 1908 and 1926 the property paid $13,474,827 in dividends.[40] A mine does not last indefinitely, however, and production at El Tigre rapidly declined during the 1930s. By 1942 the mine was idle, its dumps and drifts picked by *gambusinos*, and in 1946 the Lucky Tiger–Combination Gold Mining Company was listed as inactive in the *Mines Handbook*.

FIVE

THE BIGGEST LITTLE MINE

AS A merchant, civic leader, and banker Lemuel Shattuck is inseparably linked to the economic development of the town of Bisbee. It is a mine — the Shattuck-Arizona — that roots Lemuel's name to mining history. Development of that and other associated mines, however, resulted from a series of events that began in the late 1880s.

The last decade of the 19th Century was a momentous time for Bisbee. For three years, beginning in 1887, the shallow copper-bearing lenses of the Mule Mountains produced no less than a million pounds of copper a month, extracted through the Czar shaft of the Copper Queen Mining Company. Such production transformed Bisbee from a camp of prospectors to a full-blown, proven mining district. However, deposits of copper carbonates and oxides were shallow and limited in extent; other ore reserves had to be found if the Copper Queen was to survive.

The company literally plowed its profits back into the ground, through acquisition of additional mining property. In 1887 the Goddard claim was acquired, two years later the Holbrook shaft was sunk, and in 1890 the Gardner claim was purchased and yet another shaft commenced. Exploration from these mines soon proved that ore-bearing limestone strata extended farther to the east than anyone previously thought.

While the opening of each new mine enlarged the ore reserves of the Copper Queen Mining Company, news of the discoveries swept through national and international mining circles. Speculators and promoters descended on Bisbee like predators, hunting down impoverished prospectors and offering trifles for their claims. The rush to get in on what was perceived to be a bonanza resulted in the Copper Queen losing a number of claims which later proved exceedingly rich, among them the Irish Mag claim.

James Daly, a pugnacious Irishman, disliked Copper Queen management. His hatred, in all probability, was rooted in the fact that during its acquisition of property, the company had virtually isolated Daly's claims. With right-of-way blocked, the Irishman could do little development work at his Irish Mag claim (named after a girl in Bisbee's tenderloin). Copper Queen Superintendent Ben Williams repeatedly offered to purchase Daly's property, but all offers, no matter how liberal, were rejected. The dispute grew hotter, erupting in a fracas in which Daly was wounded by town constable Dan Simons, and incarcerated for a year in the Territorial Prison at Yuma. Upon his release, Daly declared he would never again submit to arrest and swore to get even with the "Copper Queen Crowd," particularly Ben Williams.

At this point, Dr. James Douglas attempted to conciliate Daly by again extending an offer to purchase the Irish Mag claim. The miner, however, threatened to kill both Douglas and Williams should either man set foot on his property. Shortly thereafter the Irishman reversed his stance and offered the Copper Queen not only the Irish Mag claim, but the Senator group of claims, and a claim on Sacramento Hill, all for $10,000. Douglas would have snapped up the offer had not Ben Williams, who viewed acceptance of Daly's offer as intimidation, threatened to resign if the deal was consummated.

Before Douglas could mediate the squabble, the miner assaulted a Mexican employee of the Copper Queen Mine, and shot and killed constable W. W. Lowther, who attempted to arrest him on April 10, 1890. Daly fled Bisbee and floated about the country, signing away his claims to anyone with a few dollars. On September 2, 1890, he conveyed the Irish Mag claim to saloon owner

A. J. Mehan, who in turn lost the claim through an order of execution secured by Adolph Cohn of Tombstone, to whom Mehan owed $300. That was the beginning of legal battles that would ensnare the Irish Mag claim in nine years of litigation.

Yet a third claimant of the Irish Mag appeared: a woman with an eighteen-year-old son arrived from Leadville, Colorado, professing to be Daly's wife. Her claim to the mining property was weakly pursued and she disappeared, only to be replaced by Daly's common-law wife, Angela Diaz, who presented a stronger case for ownership of the property. She had lived with Daly for five years, and advanced him money to secure title to the claim and to finance annual assessment work required by law. Upon Daly's departure, Angela had fallen on hard times, and for $1,800 signed over to Martin Costello, a Tombstone saloon keeper, what she considered her lawful property. Costello immediately contested Cohn's claim to the Irish Mag in District Court on grounds that the equitable and prior interests of Angela Diaz were well known throughout the copper camp. Although the case was appealed to the Supreme Court, Costello's title to Daly's claims was upheld on May 15, 1899.

Copper Queen management long knew the value of the Irish Mag but was hesitant to open negotiations for the property due to Daly's fugitive status, and the company was slow to react once the Supreme Court rested its case. Consequently, the Irish Mag slipped into the hands of capitalists who had been waiting to secure a foothold at Bisbee.

These capitalists were alerted to Bisbee's potential by "Captain" James Hoatson and his brother Tom, Scottish miners who had worked at Calumet, Michigan, and Butte, Montana. Quite a mythology has grown up around the entrance of the Hoatsons into Bisbee mining. The most reliable account is that given by the Bisbee *Daily Review* on June 17, 1909, which related that in 1898 John Graham, an old miner, appeared at Calumet with beautiful specimens of malachite and azurite from some claims at Bisbee. He had worked with James Hoatson at Butte, and sought out his old friend. Hoatson, who was employed by the Calumet & Hecla Company, was much interested in the specimens and inquired as to their source.

"There're from a claim at Bisbee that shows carbonate just as good as the Copper Queen," Graham replied.

That was enough to cause the Hoatson brothers to board the first train bound for southeastern Arizona. They spent weeks in Bisbee studying surface conditions and shallow inclines and prospects southeast of town. Most important, they examined the underground workings of the Copper Queen's Holbrook Mine, situated about 200 feet from the sidelines of the Irish Mag claim. It appeared to the Hoatsons that copper-bearing limestone formations tapped by the Holbrook probably extended to the southeast. In all probability Costello's claims sat atop those horizons. And the saloon keeper was willing to negotiate an option.

The Hoatsons returned to Calumet eager to form a development company. Raising money was the least of their worries. Both men had reputations as discerning miners who had developed a number of Montana and Michigan properties into paying mines. Their venture at Bisbee was accepted virtully without question, every shift boss of the Calumet & Hecla Mining Company, and the Hoatsons' friends, subscribing to stock in the company. The real money, however, was contributed by Charles Briggs, the leading banker of Calumet, Michigan. On November 9, 1899, the Hoatsons offered Martin Costello $500,000 for eleven claims, including the Irish Mag. A deal was struck, a company formed — the Superior and Western Development Company — equipment ordered, and a shaft commenced.

Mining in Arizona was different than in Michigan, where native copper outcropped at the surface and seams of metal were easily followed downward. At Bisbee there were few if any indications of ore. Because a thick, leached capping rested upon Carboniferous limestone, shaft sinking was slow and costly work. Exploration was hit or miss. The ground was penetrated 700 feet and only a few stringers of copper carbonate were detected. Drained of money, the Hoatsons shut down the operation, and returned to Michigan. But they had faith in the property, and succeeded in enlisting additional financial aid from Henry W. Oliver of Pittsburgh, and his friends Gordon R. Campbell, Chester A. Congdon and George E. Tener, all of Congdon and D'Autremont Steel, a subsidiary of U.S. Steel. The Superior and Western Development Company was reorganized in March 1901 as the Calumet and Arizona Mining Company, and work was resumed on the Irish Mag claim. A rich ore mass was encountered at the 720-foot level, followed by discoveries of bonanza grade deposits. One immense ore body, 4,500 by 1,500 feet, containing oxides and carbonates running as high as thirty percent copper, was blocked out. The Calumet and Arizona Mining Company became Bisbee's second-ranked producer. Its prosperity assured, the company acquired land at Douglas and built a smelter not far from the Copper Queen reduction works.[1]

The Irish Mag Mine was rich, but the 160 acres of Calumet & Arizona property contained limited ore reserves. With assets of $6.5 million dollars the financers moved to correct the situation. They established five other development companies, and sought out all available mining properties lying contiguous to the Irish Mag Mine. As owners of some of the most promising claims, Lemuel Shattuck, Edward Grenfell, and John Keating were among the first miners contacted by Calumet and Arizona officials.

Long before the Hoatsons appeared on the Bisbee scene, Shattuck and his associates had surmised that copper-bearing Carboniferous limestone formations extended to the south and east of the proven ground of the Copper Queen Mines. As early as 1893 they began acquiring mining properties. A quarter of a mile southeast of Sacramento Hill, the 300-foot gossan blossom that is the keystone to the semicircular copper measures of the Bisbee basin, Lemuel, his father-in-law Edward Grenfell, and John Keating held title to not only the World's Fair Group of claims, considered among Bisbee's best prospects,[2] but also the Cochise and Esperanza groups, as well as the Junction, Pay Day, Comet, Come-by-Chance, and Primrose claims. In partnership with Martin Costello, and prospectors Wallace Dwyer, George Clark, and Pete McCoy, Shattuck also had interests in such adjacent claims as the Julie, Waddell, and the Last Rose of Summer.[3]

Except for the World's Fair Group, which was somewhat developed, most of these claims showed little if any mineralization on the surface. Walking over the ground, a person would observe only a pebbly, sterile outcrop known as Glance conglomerate, a formation that would take on new meaning with the striking of copper in the Irish Mag.

In 1902 H. H. Hovland, of Duluth, Minnesota, representing Charles Briggs, Gordon Campbell, the Hoatsons, and other principals of the C&A Mining Company, opened negotiations to option thirty-nine claims, including the majority of Shattuck, Grenfell, and Keating holdings. The price was rumored to be in excess of half a million dollars.[4]

With some justification, Bisbeeites were suspicious of eastern financiers, whom they thought were out to steal their properties. It took months to get the owners together and into a frame of mind whereby meaningful negotiations could be carried on with eastern and midwestern financiers. To accomplish this, Hovland relied upon the influence of Lemuel Shattuck, who held the confidence of Bisbeeites as well as the esteem of entrepreneurs. After weeks of haggling a deal was struck on July 3, 1903. Eighteen claims — eight of which were owned by Shattuck, Grenfell, and Keating — comprising 185 acres, were bonded for $500,000.[5] Ten percent of this sum was paid down, another ten percent would be forthcoming in eighteen months, and the balance was payable in thirty months. John Keating would not benefit from the sale of his claims, however. Within weeks of closure of the mining deal, he contracted typhoid and died, much to the sorrow of Shattuck and Grenfell. They had been more than close friends; they were family, Keating and Shattuck having both married Grenfell girls.

The Junction Development Company — actually an adjunct of the Calumet & Arizona Mining Company — was organized and capitalized for $750,000 to carry out exploration on the claims. Development of the property began at once under supervision of C&A engineers already at Bisbee. The ground was probed with Sullivan Diamond drills, the first used in the district, which revealed two ore bodies at a depth of 820 feet. Less than a thousand

This photo, taken about 1896, shows a youthful Lemuel Shattuck, right top row. Center front from left: Will Grenfell, Lemuel's brother-in-law, Maurice Denn center. Young man at right is Albert (Bert) Grenfell. The older man standing next to Shattuck may be John Keating.

(Fathauer collection photo)

feet northeast of the Copper Queen's Lowell shaft, a four-compartment shaft was sunk, and a power plant and double hoist installed. By fall 1903 the Junction shaft was 400 feet deep, penetrating strata showing low-grade carbonates with a little chalcocite and chalcopyrite, copper sulfides. On October 3, 1903, the first of many large ore bodies was struck. Junction Development had become a mine.[6] For eighteen months exploration was rapidly pushed and the mine brought into production. In October 1905 Junction Development Company was reorganized as the Junction Mining Company. Stock of the shareholders of the original company was exchanged at a ratio of three shares of new stock to one of the old, and an additional 100,000 shares were issued, Calumet management taking it all. The company was capitalized at $3,000,000.

Success of the Junction Mine, as well as other C&A holdings, collectively called the Bonanza Circle Mines, whetted the appetites of other midwesterners. In 1904 W. B. Mershon, Wait Humphrey, and F. L. Harrington, who were Bay City, Michigan, lumber and mining men, purchased for $350,000 seventeen claims comprising the World's Fair Group of Shattuck and Grenfell, and a series of claims belonging to Joseph Muheim and his uncle, Frank Dubacher; in all about 297 acres.[7]

Proceeds from the sale of the World's Fair claims allowed Edward Grenfell to escape from the drudgery of mining. "Determined to pass his remaining years in quiet," Shattuck's father-in-law moved his family to San Diego. But retirement was to elude him. Shortly after Grenfell's arrival in California, he sickened and died on August 7, 1904.[8] His body was brought back to Bisbee by Lemuel and Isabella, and today the old miner rests in the family plot at the Evergreen Cemetery.

Organized in March 1904 and financed through the sale of a million dollars worth of stock, this mining venture, known as the Saginaw Development Company, tested the property with diamond drills before commencing a three-compartment shaft on the World's Fair claim.[9] Promising indications of ore were encountered, but exploration was cut short when a considerable flow of water was struck at the 910-foot level. Pumps were installed, and sinking resumed. Exploratory tunneling called drifting followed and by 1906 several ore bodies had been blocked out. On March 7, 1906, the company was combined with the American Development Company, and both reorganized as the American-Saginaw.[10] Capitalization was increased by another million dollars, most of the stock being taken by Charles d'Autremont, Jr., and others involved in the Calumet & Arizona Mining Company.

News of the activities and success of the Calumet & Arizona Mining Company, and its associated development companies, spread throughout the lumber and mining regions of the upper Midwest, bringing to Bisbee yet another horde of speculators.[11] Among them was Martin Pattison, three-time mayor of Superior, Wisconsin, and an owner of iron mines in northern Minnesota.[12] An interest in the Calumet and Hecla Development Company drew Pattison to the district, and he spent considerable time in January 1902 examining claims and talking with knowledgeable Bisbeeites regarding availability of mining property.[13]

To Pattison's surprise everyone in Bisbee seemed to have claims touted as lode bearing. All were for sale. Unfamiliar with the district's geology, he could not evaluate the prospects, and engineers and personnel of the Copper Queen and Calumet & Arizona companies were anything but cooperative. He turned to Lemuel Shattuck, who had a reputation for honesty and a hard-won knowledge of the district's geology. Shattuck in turn introduced Pattison to Martin Costello and Joseph Muheim, owners of promising claims. And Shattuck guided Pattison to Maurice Denn, a New Englander who had been a Bisbee resident since 1881.[14]

An experienced hardrock miner, and ex-foreman of the South Bisbee Mining Company, Denn "owned enough property to train the Kaiser's army on," Shattuck laughingly told Pattison, as he piloted the midwesterner to the New England Kitchen, the best restaurant in camp, where they were to dine with the other men.

At dinner Shattuck and Joseph Muheim were affable, freely imparting their knowledge of Bisbee mining conditions. Portly

Maurice Denn, who enjoyed eating as much as prospecting, expressed the opinion that the region south and east of Sacramento Hill would some day produce as much copper as Queen Hill. Costello boasted that his Irish Mag claim was well on the way to proving that point. There was something slippery about Costello, which Pattison attributed to the saloon keeper's ability to negotiate sizeable options. Without a doubt, Costello had enough confidence to wring the last dime from any man bent on investing in Bisbee mining property.

Unlike his associates at the table, Martin Pattison was a man who liked his meat well done, recalled Lemuel Shattuck. The steaks served that night were rare, much to the delight of the Bisbeeites, but to the disgruntlement of Pattison, who sent his meat back to be grilled longer. Knowing that the cook would take Pattison's rejection personally, Lemuel excused himself and proceeded to the kitchen, arriving in time to see the chef, who was always temperamental, throw the steak on the floor and tromp on it before placing it on the grill.

Shattuck retreated to the dining room, and when the steak was served again, Lemuel watched Pattison eat it and inquired as to its flavor. Pattison assured Shattuck that it was the best steak he had ever eaten. For Pattison, that dinner was the beginning of a long-standing business relationship. While Shattuck always recognized the significance of that dinner, the antics of an irate cook would be the highlight of the evening. [15]

Over the course of the next two years, Pattison became a regular visitor to the Warren District. His knowledge of Bisbee's geology grew until he was able to weed out claims as a farmer culled thistles from a field. He cast aside any notion of investing in Calumet & Hecla property and set his eyes instead on two groups of claims. The first group comprised thirteen patented claims owned by Maurice Denn, Joseph Muheim, John M. Johnson, and Lemuel Shattuck. Situated immediately east of Superior & Pittsburgh Development Company ground, this property probably sat atop the same limestone horizons that had made the Junction Mine so rich, the only drawback being that the dip of its strata would place ore at considerable depth, necessitating costly shaft sinking. The second group of claims that held Pattison's interest was located in the northeastern portion of the district, just north of the Pittsburgh & Duluth Development Company, about half a mile south of the original Copper Queen Mine, close to the old Higgins mine. Six in number, they covered a small area of only 109 acres.

Located by Lemuel Shattuck in 1893, the claims were in a canyon high on Mt. Martin or Queen Hill (sometimes called Bucky O'Neill Hill). Their access was difficult, but the geology was right. Carboniferous limestone, the prevailing rock on the claims, was cut by porphyry dikes. The most promising indication of ore, however, was a rusty colored silicious iron capping, the result of millions of years of oxidation. Faults scarred the claims: possible conduits for deposition of copper into the limestone layers, Shattuck quickly pointed out.

In light of exploration by the district's two major companies, the Shattuck and Denn properties had enormous potential. Pattison's capital was limited, however. He postponed negotiations for the Denn claims, and moved in 1903 to acquire the Shattuck property and the Leo and Roy claims, owned by Martin Costello.[16] These claims were bonded for $1,050,000, raised by Archibald M. Chisholm of Duluth, Minnesota, who made a fortune mining iron in the Mesabi Range,[17] and two lumbermen from Ashland, Wisconsin, Thomas Bardon and A. Guthrie. Louis W. Hill, son of James J. Hill of the Great Northern Railway, also contributed a substantial sum. Of that amount, Shattuck received $850,000, Costello $200,000. The Shattuck-Arizona Mining Company was incorporated under the laws of Arizona Territory by J. M. O'Connell, Bisbee city attorney and a confidant of Lemuel Shattuck.

Capitalized at $2,500,000, divided into 250,000 shares of stock, the Shattuck-Arizona Mining Company was the most closely held mining company in the Warren District. The largest block of stock, 65,000 shares, was retained by Shattuck, lesser amounts by the midwesterners. As president Thomas Bardon presided over a board of directors consisting of Lemuel Shattuck, A. Guthrie, Martin Pattison, and Archibald M. Chisholm. Shattuck was selected as mine manager, and Byron Pattison, Martin's son, would oversee development as mine superintendent.[18] The company's general offices were in the Sellwood Building in Duluth.

Because of the small acreage of the mining enterprise, and the fact that it had not scratched the surface nor had any visible outcrop of ore, issuance of 250,000 shares of stock was viewed with considerable suspicion. "The total purchase price of the property was about $950,000, the claims being six in number, the location on the highest mountain peak in the Bisbee camp," gasped the *Iron Ore* of Ispheming, Michigan. "There are no developed ore bodies anywhere near it, a great belt of quartzite cuts through its western portion and its location renders it difficult of exploration. . . . While the newspapers have been told to say there will be no stock for sale, the very fact that it has 250,000 shares on hand is proof sufficient that the object is to sell shares and also to have enough to go around. As we say, the scheme does not look right to us, and we advise our friends to look carefully into it before parting with their money in exchange for this stock."[19]

Undaunted by critics and skeptics, the Shattuck-Arizona Mining Company hired Gus Ecker, a former engineer of the Bisbee Improvement Company, to plot the course of exploration, a formidable task. Surrounding claims possessed all tunnel and adit sites. Precipitous mountain slopes rendered access to the Shattuck claims difficult, precluding an inclined approach to ore-bearing strata. A shaft would have to be sunk. After careful surveying, a site was selected for the collar (the foundation upon which the headframe rested). Timber for a headframe and shaft lining was hauled up the mountain, along with boilers and cylinders of a power plant. Byron Pattison hired a crew of skilled drillers, timbermen, and professional miners, and with Burleigh drills purchased from the C&A Company, they went to work late in September 1904.[20]

The shaft was sunk at a breakneck speed, Pattison and his miners averaging no less than 110 feet a month. By November 19, the shaft was 226 feet deep. Lemuel Shattuck was constantly at the site inspecting the shaft; and later that month Chisholm, Bardon, Martin Pattison, and three lesser investors arrived in Bisbee to observe the work. Indications of copper oxide in the shaft and the professionalism demonstrated by Byron Pattison and Shattuck built enough confidence in the company that the directors moved to acquire the Denn property, as well as adjacent claims owned by J. J. Johnson, Joseph Muheim, William Geisenberg, and Richard Bretherton. In all, about 200 acres between the Junction and Saginaw mines were purchased or leased. Superintendent Byron Pattison secured the options and completed the deal on December 1, 1904. The price tag: $750,000.[21]

In January 1905 the Shattuck-Arizona Company drove their shaft down another 130 feet, breaking the record for shaft sinking in the Warren District. On the night of February 1, Manager Shattuck and Superintendent Pattison feted their men at the English Kitchen, the finest eatery in Bisbee. The miners fed on a lavish dinner, awash with the best of wines. Toasts were exchanged: management congratulated its men on their dedication; miners reciprocated with compliments to Pattison, Shattuck, and Ecker. "All the gentlemen had agreeable things to say," reported the *Daily Review*, "which aided to the making of one of the happiest evenings that miners, mine owners, and friends ever spent together in a social session in the history of the camp."[22]

By February 18, 1905, the Shattuck-Arizona shaft was down 515 feet, and a force of men was at work cutting a station on the 500-foot level. A drift was commenced the following week to tap an ore body that diamond drilling had detected to the east of the shaft. On March 5 the first ore was struck, a small body of sulfide assaying twelve percent. Shattuck-Arizona stock leaped to ten dollars a share, and the *Daily Review* leaked the news that a local stockbroker had a few shares of the company for sale. Over the course of the next ninety days the shaft was deepened to 700 feet. At one hundred foot intervals stations were cut, and drifts extended laterally in quest of ore bodies. Signs of heavy mineralization were everywhere, reported Shattuck miners coming off duty. It was only a matter of time before something big was penetrated. Bisbee was vibrant with anticipation.[23]

Then came the announcement. At one o'clock on the afternoon

The Shattuck-Arizona Mine late in 1904.
(Photo courtesy Bisbee Historical and Mining Museum)

of July 26, 1905, Superintendent Pattison proclaimed that "his men are now working in one of the greatest and richest bodies of copper ore ever discovered in the vicinity of Bisbee, one that will equal that first rich and extensive body of ore discovered in the Mag shaft of the C&A mine."[24]

A drift on the 700-foot level had penetrated not carbonates, oxides, or sulfides, but native copper in a decomposed limestone matrix. The mass of metal was reminiscent of Michigan deposits, where native copper was the sole product. For thirty-seven feet miners hacked and blasted out chunks of the red metal, and still the deposit continued. "The breast of the drift looked almost like solid copper," reported the *Daily Review*. On July 26 several large pieces of copper were carted into town and put on display at the Miners and Merchants Bank. "This strike will prove the mine beyond doubt," the *Review* proclaimed on July 27. "It means that another great producer will be added to the district before the end of the year." Next day Shattuck-Arizona stock jumped to twenty dollars a share.

Ore strikes became a daily occurrence at the Shattuck Mine. Most of the deposits were small carbonate bodies typical of Bisbee's mineralization. Others were gigantic. In October a crosscut on the 700-foot level penetrated a mass of cuprite (copper oxide, containing 88.8 percent copper) near the Czar fault, which was declared to be "one of the richest strikes ever made in the district." The crosscut passed through seventy feet of "good smelting ore," only to encounter a flow of water that forced suspension of drifting until pumps could be installed. By close of October 1905 three large, distinct ore bodies had been blocked out.[25]

Little ore, however, was shipped to the smelter. The Shattuck-Arizona Mine was isolated high on the mountain by surrounding properties. A tortuous road, in many places nearing a twenty percent grade, led to the mine's collar, and precluded transportation of ore to the railroad spur below. The company chose not to modify the road for hauling ore, but to sink its shaft to 935 feet, enlarge it to three compartments and, for reasons of safety, connect the workings with the Cuprite Shaft of the Copper Queen Company.

A new headframe was erected, the steam plant enlarged to 600 horsepower, and an Allis-Chalmers duplex double-drum hoist installed. Working under 125 pounds of steam pressure, the hoist was capable of handling continuously a 15,000-pound unbalanced load, at a vertical speed of 2,000 feet per minute. An aerial tramway — the only one in the Warren District — was purchased to convey ore to shipping bins on a spur of the El Paso & Southwestern railway.[26]

Installed during August 1906, the 3,200-foot-long tramway, manufactured by the Trenton Iron Works of New Jersey, was supported by fourteen towers, twelve to forty feet high, built of twelve-inch timber set in concrete foundations. Suspended from a one-inch diameter steel cable and spaced at 100-foot intervals were thirty-two buckets each capable of holding ten cubic yards of ore. Once started, the tramway worked by gravity, with full buckets coming down the twelve- to eighteen-percent grade, carrying the empty containers uphill to the mine.[27] Running at its fastest speed, the tramway could move 500 tons a day, dumping a bucket every forty seconds. The optimum speed, however, was a bucket every two minutes, or about 300 tons a day. The tramway made its trial run the last week of August 1906, transporting 200 tons of ore daily, averaging twenty-five percent copper.[28]

By fall of 1906 the Shattuck-Arizona Mine had blocked out three enormous, high-grade ore bodies, and had sampled numerous lesser deposits. In places the ore was nearly solid copper, carrying high values in gold and silver. All the usual carbonates and oxides were present in magnificent formation. Crystalline reniform masses of azurite, malachite, and associated brochantite matched anything taken from the Copper Queen mines. On the 200-foot level a hydrous copper silicate was encountered that occurred as light blue radiating masses of pseudomorphs after malachite, as well as large compact masses associated with quartz and calcite. Having distinct properties, this new and rare mineral was named shattuckite and Bisbee became its type locality.[29] Huge masses of copper sulfide — bornite, chalcocite, and chalcopyrite — averaging nine percent tenor of copper showed up at depth. There was enough ore known to enable the company to produce 1,500 tons per day for months to come, if it had a way of getting that kind of output to smelter. Richly endowed by nature, superbly equipped by competent mining men who had the means to gamble, the Shattuck-Arizona was the envy of Bisbee. When cornered by an eastern newspaper correspondent and questioned about the richness of the property, Lemuel Shattuck swelled with pride and proclaimed that the "Biggest Little Mine" sat upon his property.

The mine was rich and well engineered, but would be conservatively run. In September 1906 its production got off to a modest start of under 100 tons per day, only to be jeopardized on the night of October 13 by a fire that destroyed the new woodframe office building and the two-story blacksmith and machine shop. Fortunately, the blaze was discovered in time and extinguished before it spread to the shaft and power plant. The mine suffered no curtailment of output, and the destroyed buildings were rapidly replaced with more fire resistant plaster structures.[30]

While exploration continued, the company steadily increased production toward a goal of 300 tons per day. By January 1907, 160 men were at work, and another large ore body was tapped between the 700- and 800-foot levels. Throughout this period of discovery Lemuel Shattuck went daily to the mine to inspect ore faces, and to collect specimens and supervise their transportation into town. Beautiful specimens of crystallized native copper, azurite, and malachite adorned the lobbies of Miners and Merchants Bank and the brokerage firm of Duey and Overlock in the Muheim building on Brewery Gulch, which was handling the sale of small blocks of Shattuck stock that somehow mysteriously appeared. Shattuck never left the mine without stopping to chat with his employees, many of whom he knew personally. He inquired as to their health and the well-being of their families, and he sought their opinions regarding direction and complexity of ore streaks. Regular visits to the mines was a pattern Lemuel Shattuck maintained almost to the day he died.

The *Iron Ore* was correct in its assumption that the large number of shares of Shattuck-Arizona stock was due to the desire

The Shattuck-Arizona Mining Company tramway, the only overhead ore haulage system ever built at Bisbee. Sacramento Hill at right center.
(Fathauer collection photo)

of its management to place large blocks on the market. But Mother Nature proved the correspondent wrong in his insinuation that the mine was a scam to bilk investors of their money. There was indeed ore in the Shattuck-Arizona Mine and, to publicize that fact, Lemuel and other principals of the company used every means at their disposal: from collecting and displaying its minerals to publishing news releases detailing every strike on the property. The company's stock soared with each strike of ore. Within a few months Shattuck stock was considered among the best mining investments, along with that of Superior & Pittsburgh, which owned the Junction Mine.

In January 1907 the price of copper hit twenty-five cents a pound, sending mining stocks sky high. Stock in the Junction Mine broke a hundred dollars a share. At fifty to sixty dollars a share, Shattuck-Arizona stock was a bargain, with a degree of glamor attached to it. Carl H. Wiel, a wholesale liquor dealer of Chicago who had mine holdings in Arizona, presented his bride, the daughter of a prominent Spokane banker, a gold horseshoe that was set with thirty half-carat diamonds, and 2,000 shares of Shattuck stock with a market value of $54 a share, totaling $108,000.[31] The wealthy had faith in the mine, and so did its management — faith enough to attempt to duplicate the success of the Bonanza Circle people.

In late January 1905, before the Shattuck-Arizona produced ore, its management commenced development of its property between the Junction and Saginaw mines, near the Evergreen Cemetery. The Denn-Arizona Development Company was established, with 75,000 shares of stock at $10 par, taken entirely by men financing the Shattuck.[32] Diamond drilling began at once. For two months formations were tested to determine if there was any mineralization and at what depth the ore lay. When a core revealed disseminated cuprite averaging twelve percent copper, at a depth of 950 feet in the direction of the Saginaw Mine, the next step was ordered.[33]

Sinking a shaft required competent engineers and skilled miners, men who were in short supply due to exploration then underway in the Warren District. Byron Pattison feared that a competent foreman could not be found to head the underground work, and that months would be wasted searching for such a man in other districts.

As a banker, miner, and merchant, Lemuel Shattuck knew every hardrock miner having supervisory ability. While he conceded that a good engineer might be hard to find, he assured Byron Pattison that if one existed in Bisbee he would locate him. Out of the proverbial hat, Shattuck pulled not a rabbit but Joseph Collie, who had engineered a number of shafts for Bonanza Circle development companies. Not only could he sink a straight shaft, he knew the formations at the southeastern end of the Warren District.[34]

When a three compartment shaft was commenced in March 1905, the company's stock, valued at only thirty cents a share, rose to $2.70 a share. It would be months before the stock took another jump. Sinking the Denn shaft was slow, tedious work — this time no records would be set. Collie and his men drilled and blasted their way through the tough conglomerate capping, and at a depth of 840 feet passed into limestone. The shaft was sunk further. On the 900-foot level stringers of ore were encountered. A second shaft was started in May 1906 on the Lee claim, while the original shaft was deepened another 100 feet. Three months were spent driving a drift to connect the two bores. Leached ore was encountered, but nothing like what was showing up in the Shattuck Mine across the gulch. Denn stock slowly advanced with purchasers' anticipation of finding ore.[35]

At a distance of 525 feet from the shaft, Collie ordered a south-trending crosscut driven from the drift to bisect the area where diamond drilling had exposed ore. On December 10, 1906, a round of dynamite charges set off in the face of the crosscut exposed the "good stuff" — a face of red, earthy cuprite. Eager to see if the ore had substantial dimension, the miners drove the crosscut another twenty-five feet. Solid ore all the way.[36]

This striking of a sizable ore body on the extreme eastern boundary of the Warren District was of great importance. It proved not only Denn-Arizona ground, but that of the American-Saginaw and other developments closer to Warren. Great mining companies are built upon such news. On December 10 Denn-Arizona stock stood at nineteen dollars a share. Shots fired that morning in the mine set off the fireworks. Crowds filled the brokerage firm of Duey and Overlock. There was only one topic of conversation in Bisbee: Denn-Arizona and Denn-Arizona. So long as the full breast of red copper oxide continued so did the upward spiral of Denn stock. In three days the stock doubled in value.

Continuation of the crosscut showed up more ore. A round of shots on December 27 exposed yet another "full face" of thirteen percent ore. On the basis of the two finds, the shaft was sunk another 100 feet and a drift run directly beneath the one on the 1,000-foot level. By August 1907 a seventy-foot extension of ore had been blocked out between the drifts, proving that the ore had depth.

Because the Denn was a deep mine, control of water became a problem that nagged at exploration and production. What started as seepage at a few hundred feet increased to a flow of hundreds of gallons by the time the 1,000-foot level was reached. Two 1,000-gallon triple-expansion Prescott pumps were installed at the 1,000-foot level. With water flow somewhat controlled the shaft was sunk deeper in search of ore. Stations were cut and drifts and crosscuts run. While stringers of low grade sulfide were encountered at 1,250 feet, rich carbonate showed up only sporadically. Management felt, however, that the mine contained enough ore to justify reorganization into a full-blown mining company. On January 14, 1907, the company was reorganized as the Denn-Arizona Mining Company, capitalized at slightly more than three million dollars, with 300,000 shares of stock at ten dollars par. The same men that sat on the board of the Shattuck-Arizona controlled the Denn: Martin Pattison, Thomas Bardon, Archibald Chisholm, and A. Guthrie. Maurice Denn and Lemuel Shattuck were directors as well.[37]

Besides the Shattuck and Denn Mines, Lemuel Shattuck had an interest in the Cochise Copper Mining Company, whose claims were located in Dubacher Canyon, close to the Denn and Saginaw developments. Consisting of fifteen patented claims comprising 175 acres, this company was organized on March 19, 1898, by Shattuck, Joseph Muheim, Lee Ross, J. H. Mathias, C. L. Jones, Charles B. Watts, and Charles Forcade. By July 1900 three shallow shafts had been sunk and $35,000 expended in search of ore, with negligible results. Depleted of running capital, the Cochise Copper Mining Company lay idle until development of the Junction Mine and exploration at the Saginaw and Denn revived Shattuck's interest in summer of 1905.[38]

With proven ore horizons at the southeastern end of the district, Shattuck and his colleagues decided to gamble on the Cochise claims, organizing on August 1, 1905, the Cochise Development Company. Incorporated in Arizona, the enterprise was capitalized for $1,000,000, with 100,000 shares of stock at $10 par. Sixty thousand shares of stock were eventually sold, mostly to Bisbeeites, with Joseph Muheim (12,031 shares) and Lemuel Shattuck (4,407 shares) the largest shareholders.[39]

By fall of 1905 a new road had been completed to the shaft site, and a headframe, engine and boiler house, blacksmith shop, and framing shed erected. A first motion, double-drum hoisting engine, capable of handling work at a depth of 1,200 feet, was installed, as were a straight line Sullivan air compressor with a capacity of six drills, and a 100-horsepower Atlas boiler. The mine was equipped with Ingersol drills, ore cars, buckets, and necessary mining supplies.[40]

Late in 1905 Shattuck hired thirty-seven men and put them to work in three shifts. They pumped the old shaft dry, enlarged it to three compartments, and deepened it 110 feet through ground that Lemuel declared "the hardest in the camp." Mineralization the full extent of the shaft drove the miners on. By May 1906 they were drifting westward under Chihuahua Hill, encountering occasional stringers of ore showing fair value in copper with some gold and silver.[41]

In September the 300-foot level drift penetrated a succession of talc slips containing copper sulfide assaying as high as twenty-two percent. News of the discovery brought Lemuel to the mine, and he collected specimens of black sulfide for exhibition at Duey and Overlock, and at the Miners and Merchants Bank. Word of the strike brought a rush of people seeking Cochise stock to the brokerage firm, but none was available.

As the talc slips dipped to the north and west, Lemuel believed an ore body probably existed up the canyon from the shaft site, and ordered drifting in that direction from the 600-foot level, as well as to the southeast from the 900-foot level. Exploration continued until September 1, 1907, with disappointing results. The ground was mineralized, but barren of commercial quantities of ore. Barren too was the company's treasury, and Lemuel discontinued work. [42]

Shattuck and the Muheims nevertheless had faith in the Cochise claims, believing ore bodies existed at depth just as at the Denn Mine. They paid taxes on the property over the years, and in 1914 leased the claims to I. N. Kinsey of Douglas, who also optioned to buy the property for $350,000. When Kinsey failed to conclude the option three years later, Shattuck and the Muheims offered the property to the Copper Queen Consolidated Mining Company for $500,000. Nothing came of that deal except leasing of a portion of the land to the latter company for construction of a switchback and railway for its Sacramento Pit operation. Disappointing as the Cochise claims were, Shattuck and the Muheims never relinquished the property, and at a far later date the Cochise claims would figure in a revival of mining at Bisbee. [43]

Although the Shattuck-Arizona was shipping only 200 tons of ore daily over its tramway by September 1906, its workings contained a vast reserve of copper, enough to keep furnaces of a reduction works going for years. Then too there was every prospect that the Denn Mine would pour forth ore. For the moment, the limited production was forwarded to the Copper Queen smelter at Douglas for refining. Although the high cost of custom smelting was somewhat offset by the tenor of the ore and the high price of copper, the expense of refining would nevertheless cut deeply into profits over the long run. It was only natural that Shattuck directors began to toy with the idea of smelting their ore.

In mid-November 1906 the directors of the Shattuck-Arizona Mining Company converged on Bisbee to discuss the feasibility of building a smelter. Meeting at the Miners and Merchants Bank on November 16, the Pattisons, Chisholm, Guthrie, Shattuck, and Denn poured over estimates of ore reserves and smelting costs. Opinions were mixed: Shattuck, Bardon, and the Pattisons wanted a refinery. Other directors and investors were not sure. They questioned whether ore reserves, even at the current high price of copper, were adequate to justify the cost.

They all concurred on several points. Should the company decide to build a smelter, land with artesian water was available at Douglas, twenty-five miles south of Bisbee, near the Mexican border. And word of a contemplated smelter would surely drive up company stock. In fall of 1906 the Shattuck-Arizona Company optioned a 250-acre tract between the Copper Queen and Calumet and Arizona smelters. While there was disagreement regarding the merits of a refinery, Shattuck directors unanimously voted to announce plans to erect a smelter. On November 17, 1906, the *Daily Review* broke the news that the "Shattuck-Arizona plans to smelt their own ore." As expected the company stock soared, by twenty dollars a share, and Denn stock threatened to break eighteen dollars a share. In the meantime, the directors would study the feasibility of the project. A wise decision, in light of strong indications by January 1907 of a weakening commodities market.

SIX

THE SHATTUCK JUGGLING ACT

BY THE time he was forty years of age Lemuel Shattuck was managing three large-scale mining enterprises, a bank, a lumberyard, rental properties, a saloon with all its attendant headaches, and a keg beer distributorship. Such enterprise would have burned out most men. Not Lemuel. He thrived on business and sought more, and at the same time lovingly looked after an expanding family and fulfilled civic responsibilities, which included serving several years as a councilman for the town's third ward, and sitting on every board formed for town betterment. It was a juggling act in the finest sense of the word, in that Shattuck dropped none of the balls.

Besides mining interests, Lemuel had enterprises beyond Bisbee, at Douglas in the Sulphur Springs Valley — the result of fast moving competition against the "Copper Queen Crowd." This story, worthy of a book in itself, started in 1900 when James Douglas, developer of the Copper Queen Consolidated Mining Company, sought to move his smelter from its cramped canyon site at Bisbee to the lower Sulphur Springs Valley, where unlimited quantities of water existed below the surface. The twenty-five miles separating the mines and the new refinery site would be spanned by an extension of the Arizona and Southwestern Railroad, which would link at the border with the Nacozari Railroad.

Thus the refinery would receive ore from mines both in Bisbee and Mexico.

Of course, any plan for a smelter would include a townsite and public utilities, and in October 1900 Douglas, William and Michael Brophy, Michael J. Cunningham, Ben and Louis Williams, and others affiliated with Copper Queen management, as well as John Slaughter, incorporated the International Land and Improvement Company to build a town, appropriately called Douglas. Because there was every prospect of enormous financial gain from this enterprise, no fanfare accompanied the organization of the company. That would change the minute survey crews went to work. In December a 160-acre tract on Whitewater Draw, twenty-five miles southeast of Bisbee, was surveyed and staked. As a result of slowness on the part of Michael Cunningham to record the preemption with the Federal Land Office at Tucson, another group intervened in the scheme.

While in the Sulphur Springs Valley buying cattle, Charles Overlock heard from H. H. Whitney, an old associate in the Erie Cattle Company, that surveyors were staking off a section of land near the Mexican border. Overlock immediately contacted associates in Bisbee: among them his old friend Lemuel Shattuck and real estate promoter S. K. Williams. They correctly guessed what the Copper Queen Crowd was up to and promptly threw a "monkey wrench" into the works. Inviting in J. E. Brock, H. H. Whitney and Alfred Paul, they formed a rival company called the Erie Townsite, and Overlock went to Tucson to preempt the 160 acres the Copper Queen Crowd had staked, but not recorded. From his lumberyard, Lemuel delivered fence posts and barbed wire, and his crew began fencing the ground. George C. Clark surveyed the property and split it into lots. Real estate promoter S. K. Williams began advertising the Erie Townsite.[1]

Dismayed by the fast moving Erie people, Cunningham attempted to correct his oversight by filing on an adjacent section of land. But the damage had been done; Douglas and his associates had been upstaged by Overlock and Shattuck. With two townsites there was slim prospect of anyone making money, an explosive situation to say the least. Sound business sense prevailed, however.

Late in January 1901 the International Land and Improvement Company invited the Overlock-Shattuck party to join the townsite company. The compromise was accepted and the Erie Townsite was taken off the market.[2] Although the "jumping" of the Douglas townsite created momentary hard feelings, it was probably a good thing. For the inclusion of Shattuck and Overlock in the townsite development broke the Copper Queen monopoly, and brought into community planning a wealthy and powerful group that rapidly developed Douglas.[3]

In partnership with Joseph Muheim, Charles Overlock, and Peter Johnson, Lemuel began a building program in the heart of the townsite. In 1901 they shipped in enough brick from the Arizona Clay & Manufacturing Company of Benson to build a two-story commercial building at the corner of 8th and G Streets.[4] In spring 1902 Lemuel and Charles Overlock founded the Douglas Lumber Company, which operated in both a retail and wholesale capacity. With five million feet of lumber and all kinds of building materials, its yard at 6th and I Streets dwarfed the Shattuck Lumber Company at Bisbee. In characteristic fashion, Lemuel delegated duties. Charles Overlock assumed the managership and became company figurehead. The two men also erected a three-story hotel at 10th and H Streets in fall 1902.[5] Four years later, Lemuel, the Overlocks, and Muheim, along with the Bishop of the Mormon colony of Colonia Dublan, seriously considered investing in 120,000 acres of timberland in the Sierra Madre of Mexico, building a railroad to transport logs to a projected sawmill at Douglas.[6]

As at Bisbee, Lemuel Shattuck quietly moved in the background of Douglas commercial activities, financing and building. Not widely known is that Shattuck helped finance the building of the world-famous Gadsden Hotel, for years the showplace of Phelps Dodge power in the smelter city. In 1905 he assumed a first mortgage on the edifice, which he carried for decades through good times and bad.

Lemuel constantly worked to improve his properties at Bisbee.

Sitting on lower Brewery Gulch, the St. Louis Beer Hall had often been inundated by floodwater that swept down the defile every summer. During such deluges Shattuck, and every other proprietor along the town's two thoroughfares, made no attempt to divert or restrain the swirling water with sandbags or barricades. They merely opened the doors and let the water flow through the buildings. Once the flood had subsided, hours would be spent removing mud and debris. After such occasions Shattuck longed for a building with a higher foundation.

By 1903 he had profited enough from his enterprises and mining ventures to realize that dream. There existed a piece of land between the St. Louis Beer Hall and the Red Light Saloon, owned by Shattuck's long-time friend and business associate, Jacob Schmid.[7] That summer the two men pooled their resources and laid plans for a two-story brick building that would house the saloon on the ground floor and contain a suite of rental offices on the second floor. Cost of the edifice was estimated at $7,500.[8] By the time construction began in July 1903, Shattuck and Schmid had altered their plans to include a third story, increasing the estimate of construction by nearly $3,000. Bisbee would have a "sky scraper," trumpeted the *Daily Review*, the third story being reserved for apartments and a private club.

Just what kind of club the paper did not explain. In all likelihood, Shattuck and Schmid envisioned the type of male retreat popular in western mining towns feeling the swells of temperance and anti-gambling movements. A portion of the third floor would be set aside where Bisbee's professional men could relax, converse, smoke, drink, gamble, and transact business away from prying eyes of do-gooders and reformers, and jostling of drunken miners and cowboys.

Like everything Shattuck set his hand to, this building reflected permanence. The foundation lay on bedrock, twenty-five feet below the surface of Brewery Gulch, providing a spacious basement for beer and liquor storage. Its outer walls were of fire-resistant red brick, obtained from the Arizona Clay and Manufacturing Company at a cost of twelve dollars per thousand.[9] A facade of trimmed rock decorated the walls facing the street, and a concrete slab bearing the date of completion—1904—capped the trim. It was, and is, one of Bisbee's finest examples of architecture.

The building's offices were quickly rented. J. A. Karlson, whose tailor shop was formerly on Main Street, opened a showroom on the second floor. Assayer and mining engineer George C. Clark rented an office, as did Bisbee City Attorney J. M. O'Connell. Since these men had close ties to Lemuel, consolidation of their offices under one roof no doubt had its advantages. The Shattuck-Schmid building, like the Miners and Merchants Bank, was a bastion of power.

Within a few days of the building's completion in January 1904, Lemuel opened his new saloon. He dropped the former name St. Louis Beer Hall and, befitting its spacious quarters, renamed the establishment "The Shattuck." The drinkery sparkled. A new, polished mahogany bar, boasting a brass footrail, ranked with the longest in camp. Mirrors and chandeliers reflected the glint of hundreds of glasses, and its bartenders, dressed in white aprons, worked three shifts seven days a week. Dealers at the gaming tables wore suits. Saloon policies remained unchanged. Anheuser-Busch beer was still dispensed—on both a retail and wholesale basis—and Squirrel Whiskey was added to the line of hard liquor. Women were never allowed on the premises, and rowdiness never tolerated. Games were honest and the booze genuine at two shots for twenty-five cents. A customer was never shorted. If a man could not finish the second drink, he was issued a token redeemable for another drink at a later date. A buffet of snacks was always available for patrons, and a traditional turkey meal was served every Christmas.

The building at 12 Brewery Gulch that formerly housed the St. Louis Beer Hall was taken over by Sid Harris and George Roberts. These veteran mixologists remodeled the interior, covered the floors with linoleum, and installed the latest bar fixtures, including a soda fountain and an improved Babcock ice box. The upstairs rooms were converted to a "summer beer garden," tastefully appointed with potted plants, antique oak tables and metal chairs.

The Shattuck-Schmid Building as it stands today.
(Westernlore Press photo)

The Hermitage Saloon, as the new resort was called, catered to a mixed clientele: men who liked alcoholic beverages as well as those with a sweet tooth, who craved soda pop. Not only were fountain drinks served, but also Zeigler's Ice Cream, the best in town. Like its competition next door, this bar ran three shifts. [10]

The most visible of Lemuel Shattuck's enterprises — the one he found the most fulfilling — was the Miners and Merchants Bank. Within two years of its founding, the institution had outgrown its old quarters in the rear of Henkel's jewelry store, forcing Lemuel to cast about for property on which to erect a new bank building. In spring 1904 the Bank purchased the Anheuser Saloon on Main Street from Otto Geisenhofer,[11] tore the structure down and built a sturdy pressed-block building but a few doors from its competitor, the Bank of Bisbee.

Keeping pace with development of Bisbee and southeastern Arizona, the Miners and Merchants Bank soon outstripped all rivals. In fall 1904 the Bank purchased property in the booming railroad and smelter town of Benson, where it intended to open an office. Because the bank charter specified only offices in the Warren District, the Benson branch was not opened.[12] A year later, however, Lemuel created a branch office in Lowell to take advantage of opportunities created by development of the Bonanza Circle mines southeast of Bisbee. That branch of the Miners and Merchants Bank, under supervision of C. A. Bennett, met the needs of businessmen in satellite communities sprouting in the Warren District, and offered savings accounts and check cashing services to hundreds of miners employed by the C&A mines. [13]

Growth of the Bank was phenomenal to say the least. By 1906 a burgeoning escrow department had been founded, which handled the sale of commercial and residential real estate. The institution, along with other Bisbee banks, financed civic improvements. When citizens voted in $80,000 worth of sewer bonds, the camp's three banks financed them in 1907 at six percent interest, four percentage points below then-current interest rates. Of the total bonds, the Miners and Merchants Bank took $40,000 worth; the Bank of Bisbee, $25,000; and the First National Bank, $15,000.

Transactions related to mining properties brought in the heavy deposits. And most of those transactions involved sale or bonding of properties that Shattuck or his friends owned or had interests in. All legal and financial matters related to the purchase of claims forming the Shattuck-Arizona, the Denn-Arizona, and the Cochise companies were administered by the Bank, as were many financial dealings involving Bonanza Circle development companies. The institution also served as site of board meetings with midwestern directors and stockholders of the Shattuck and Denn mines. The major impetus that pushed assets of the Miners and Merchants Bank beyond those of its competitors was the unusual volume of deposits in 1906 credited to the original owners of the Lucky Tiger–Combination Gold Mining Company.

Under the shrewd guidance of Lemuel Shattuck and his cashier, Thomas Hood, the Bank in March 1907 outstripped its competition by a wide margin, claiming total resources of $1,798,854. A conservative policy of management had relegated the Bank of Bisbee to third place, behind the Bank of Arizona at Prescott. While these two competitors jockeyed for second and third place positions in Arizona Territory, the Shattuck bank held the lead for years to come.

In his 1907 annual report, Territorial Auditor John Page praised the Miners and Merchants Bank, stating that its deposits "were larger totals than those presented by any other bank in the history of the territory, whether national or territorial. . . . The amounts of its deposits . . . speak well for the progress of Bisbee and the Warren District."[14]

Although the Miners and Merchants Bank excelled in attracting deposits, its record in a peripheral activity was terrible. Its baseball team, the Blue Ribbons, never measured up to the Giants fielded by the Bank of Bisbee.[15] What with "Puffy" Boswell on the pitcher's mound, "Jasper" Hood as best all-around player, Lemuel Shattuck on third base, and cousin Rawle B. "Little Duchy" Coover[16] in right field, the Bank was trounced three years in a row at the "Great Bankers Baseball Game" played on the Warren field. The Brophy-Cunningham team administered their worst defeats in July 1908 (13-2), and in August 1909 (21-10). The following year, the Blue Ribbons narrowed the gap somewhat. They were defeated 13-8. "The only trouble with our opponents," remarked one Giant, "was that each player could play like a big leaguer, every position on the team, except his own."

The year 1907 dawned on a robust American economy. Investments abounded, money was easily made in stocks and commodities. Factories and mines worked around the clock. In January the price of copper stood at an all-time high of twenty-five cents a pound. Price of the red metal slumped slightly in March. No one paid much attention; after all, fluctuations in commodity prices were to be expected. When copper plummeted to fifteen cents a pound panic set in, the worst since 1892 when large financial institutions went to the wall. Over the next several months stock values declined as much as thirty percent.

Railroad securities were the first to suffer. Millions of dollars worth of transportation stocks had been bought on margin. When the slump hit, stockholders rushed to cover their losses, selling industrials on such an enormous scale that the entire stock market was demoralized. Speculative values melted like snow in a chinook; those who believed themselves affluent found themselves poor and in debt. [17]

The financial crisis deepened across the country, affecting western mining towns by throwing thousands of men out of work. The stock of Bisbee's major producers slumped drastically. By August 18 Calumet & Arizona stock, which had stood at $198 per share, retreated to $148; Superior & Pittsburgh dropped from $29 to $13; Shattuck stock standing at a high of $55 dove to $24; and Denn-Arizona sank from $32 to $8.

On November 10 Bisbee's copper companies reduced wages by fifty cents a shift. Fearing loss of vital employees, and thinking copper would rebound in price, the large companies refrained from laying men off, reducing hours instead. [18] Because the Shattuck and Denn mines were still in developmental stage, Lemuel and Byron Pattison closed the properties, keeping only a small force of

Above: the logo of the Miners and Merchants Bank. This drawing was made by artist Ruth Bush from the copper relief mounted over the front door of the Bank building in Bisbee (below).

men to maintain the mines and continue exploration.[19] These cutbacks naturally affected other businesses in camp. Merchants found their shelves laden with goods for which there was no market, and bills coming in had to be met. Restaurant tables were vacant; miners spent little on entertainment. Bank deposits shrunk and cash reserves dried up. Loans were impossible to make or collect. One business after another closed, never to reopen.

The depression, along with revolutionary activities south of the border, squelched the Overlock-Shattuck plans to develop Sierra Madre timberland. In September the brokerage firm of Duey and Overlock failed, pulling some investors down with it.[20] Drop of the stock market would have been accepted as cause of bankruptcy, had Harry Duey not taken funds, including money from the Cochise Development Company, with which to purchase stock and then failed to deliver the securities.[21] He was not only indicted by a grand jury for fraud, but was sued by Lemuel Shattuck for embezzlement. The name Overlock was tainted by association. Lemuel Overlock sought to clear himself and his brother of wrongdoing by depositing $32,000 of his own funds into the business. Although Overlock saved his business, the firm was a long time recovering its former status. William Traux, owner of the English Kitchen, Bisbee's largest and finest restaurant, sold out in early October and moved to Sylvanite, New Mexico, where he opened the Sylvanite Cafe.[22] On March 25, 1908, the First National Bank of Bisbee closed its doors. Failure of that institution was due not so much to the financial debacle as to embezzlement of funds by its officers.[23]

Plunge of the stock market wrung cash reserves from national banking institutions. The great Knickerbocker Trust Company of New York City failed. Other eastern banks, which held assets of Arizona banks, froze accounts and halted transfers of money. To avert runs on their institutions, bankers throughout the United States adopted what was called the clearinghouse system: issuing denominational drafts instead of coin or currency. Bisbee's copper producers were unable to meet payrolls in cash.[24]

Under ordinary conditions Bisbee banks retained sufficient cash to handle withdrawals. Fortunately both the Bank of Bisbee and the Miners and Merchants Bank foresaw hard times, cut costs, and increased cash reserves. At end of summer, however, Bisbee banks were under siege, their solvency seriously threatened. Something had to be done to forestall a run on the banks, as was happening elsewhere in the country. The town's financiers first attempted to soothe townsfolk by telling them that all was well in Bisbee. Their money was safe in local banks. While the town's economy was hinged to copper, price of the metal would rebound. "Bisbee was resilient, better able to withstand the slump than other communities." Such comments were true, but not always accepted by miners who had just had their hours and wages cut.

Signs of panic loomed on Bisbee's horizon by fall. Because mining companies could not get cash from their New York banks, a meeting of local bankers was called on the night of November 4, 1907. Attending the session were W. H. Brophy, M. J. Cunningham, J. T. Hood, Lemuel Shattuck, E. W. Spiers, W. J. Eddleman, J. H. Nolan, I. W. Wallace, W. H. Rankin, and E. W. McKee, representing all banks in the city. The nation's financial condition as it affected the Warren District was the chief topic of discussion. Apparent to all was the impossibility of obtaining currency from major reserve centers, and if Bisbee did not at once adopt the clearinghouse system, cash reserves in town would soon be paid out and withdrawn from circulation, leading to a paralysis of business. The bankers unanimously voted to implement the system, and elected a board of officers to oversee the functioning of the clearinghouse: W. H. Rankin, treasurer; W. J. Eddleman, secretary; W. H. Brophy, vice president; and Lemuel Shattuck, president.[25]

The clearinghouse, under Lemuel's direction, would see that commerce went on as usual in Bisbee, except that no more than $10 in cash could be withdrawn at any one time from local banks. Instead of paying for purchases with currency or coin, a person would pay either in drafts of standard denominations drawn by the mining companies, or in certified checks, personal checks, or cashiers' checks issued by local banks. All these vouchers would be accepted in the usual course of business the same as gold coin. The mining companies would meet their payrolls in drafts of $5, $10, $20, and $50 denominations. These printed pieces of paper, called "chin-plaster" by some, would circulate the same as cash. Once the economy stabilized, the certificates could be redeemed at face value in currency.

The economy did stabilize and the clearinghouse relaxed its stance on currency. Except for some complications, such as Wells Fargo Company — the United Parcel Service of the Victorian Era — demanding payment in gold for delivery of merchandise,[26] the "chin-plasters" worked. When money flowed again at year's end, Bisbeeites exchanged their certificates for coin or currency. With restoration of currency transfers from the East, mining companies resumed paying employees in cash on January 9, 1908.

The Warren District's economy was pegged to the price of copper, which hovered under fourteen cents a pound for the next two years. Consequently mining proceeded slowly, the companies working their crews half time. Reduced wages affected banking on all levels, forcing Bisbee's institutions to cut back wherever possible. Lack of deposits forced closure of the Miners and Merchants Lowell branch on February 1, 1910, and that same day the Bank of Bisbee closed both its Lowell and Naco branches. The southern end of the district was not left without banking facilities, however. In a move that would probably be illegal under today's banking laws, the directors of Bisbee's two major banks — W. H. Brophy, M. J. Cunningham, and Lemuel Shattuck — pooled personal resources amounting to $25,000 and incorporated the Bank of Lowell in February 1910. This bank would function independently of both the Bank of Bisbee and the Miners and Merchants Bank.[27]

The Miners and Merchants Bank weathered the debacle of 1907–09, but its financial foundation was weakened by withdrawals, and its books were cluttered with loans that were uncollectable in Bisbee's current economic climate. By early 1910 Shattuck was painfully aware that his bank would fail the territorial bank examiner's audit due later in the year. Such a catastrophe he could not permit, and he chose to cover the many "lame" notes with his own resources—the most liquid of which were his 65,000 shares of the Shattuck-Arizona Copper Company.

The time was ripe for disposal of mining stock. Copper had advanced to fourteen cents a pound; the Shattuck Mine was earning more than $5,000 a day, and its stock stood at $25 a share. Early in March Shattuck informed fellow board member Thomas Bardon of his decision to bail out of the company, and empowered Fred H. Merritt, a Minneapolis metals broker, to dispose of the stock.

Stunned by Shattuck's announcement, and fearful that sale of such a large block of stock would undermine value of the company's securities, Bardon rushed to assemble the principal stockholders, and at the same time slow Shattuck. "Remember you must first get your friends together in justice to them," Bardon wrote on March 28.[28] Next day Bardon hinted of aid:

I don't want you to sacrifice your holdings in the Shattuck. You have pioneered it so long, and it now looks to me as though, through a combination of adverse circumstances which you are not personally responsible for, you are put in a position where you are being virtually forced to let go something good, to make up for some one else's mismanagement. Don't give up until Mr. Hill returns. He thinks much of the mine, and likes you personally very much. We all have a warm spot for you . . . I will wire you when to come to meet us, and we can go over the situation with you, and see how much you need to put your bank in good shape so the examiner can find no fault, and give you the relief from worry that you so much deserve.[29]

Lemuel both respected and feared his midwestern associates. They were financial sharks who would seize any opportunity to turn a profit, and the condition of the Miners and Merchants Bank made him extremely vulnerable. Now that word of his plight was out, Shattuck moved fast. In an all-or-nothing deal, he granted Merritt an option to dispose of his stock. "I do not want . . . to sell a few thousand shares and then fall down, because then Bardon and the balance [of the crowd] will give me the horse laugh, and try and handle me as they wish thereafter."[30]

Moving such a large block of stock was no easy task. The "crowd"—Bardon, Chisholm, Pattison, and Hill—were after dividends and wanted to hang on to their stock. Bisbee stockholders such as Muheim, Overlock, and O'Connell, considered the stock worth $50 per share as an investment.[31] Shattuck could control his Bisbee associates, but was powerless to block the actions of Duluth and St. Paul men, who would throw every obstacle in his path to preserve the value of Shattuck stock.

Merritt plunged into the task of moving Shattuck's stock. To increase value of the stock, and to acquaint eastern investors with a relatively unknown mining company, he published glowing accounts of the Shattuck-Arizona in eastern financial journals. Merritt placed a prospectus into the *Boston News Bureau*, the financial organ of the copper market, and succeeded in convincing several large brokerage firms to recommend the stock to their clientele. At the same time he fought to keep the price of Shattuck stock above the $25 mark. Despite herculean efforts, Merritt failed to sell the 65,000 shares in the time frame set by Lemuel, mainly because Martin Pattison had forced on the market $35,000 worth of his stock in mid-April.[32]

Nevertheless, Fred Merritt felt confident he could sell all of Lemuel's stock at $25 a share, providing the stock was split into small blocks of a thousand shares or less. But Lemuel stood firmly by his initial instructions to the broker, and canceled Merritt's option on May 8.[33] Gamely Merritt begged Shattuck to permit him to continue his efforts: "Please at least give me 10,000 shares to work on." Shattuck relented and Merritt succeeded, and on May 16 the broker telegraphed "have sold subject to your confirmation first 10,000 shares at a good price." Merritt added, "Can handle balance [55,000 shares] in 90 days." And it was a good price. After deducting Merritt's commission and expenses, Shattuck netted $250,000.

This sum was more than enough to prop up the sagging Miners and Merchants Bank. Knowing that he would receive in October a sizable dividend (a dollar a share) on his remaining 55,000 shares, Lemuel relinquished no more stock. Although the sale was made in desperation, Lemuel benefitted beyond just saving his bank. Merritt's movements in Boston focused attention of investors on the

Shattuck-Arizona Copper Company and paved the way for its listing on the Boston Stock Exchange.

The Copper Queen mines and the various Calumet & Arizona companies ran through the depression of 1907–09 at a reduced rate of production. The Shattuck-Arizona and Denn-Arizona mines, which had just begun to produce when the economic slump occurred, shut down altogether. Small crews were kept on to maintain the miles of workings and to carry on diamond drilling and exploration. During their inactivity both mines experienced some mishaps. With cessation of pumping, lower levels of the Denn-Arizona filled with water.[34] But management did not view that as a calamity, for the mine's pumps were capable of lowering the water level should the company be reactivated. The Shattuck-Arizona, however, came close to being destroyed the night of November 19, 1907.

About midnight a watchman smelled smoke coming from the shaft. He did not stop to ascertain the fire's location, but turned in the alarm immediately. With three miles of tunnels, drifts, cross-cuts, and stopes—all reinforced and buttressed with Oregon pine—a fire could be disastrous. If not brought quickly under control, it could eat its way into sulfide ore bodies and there smolder indefinitely, generating toxic gases that would render the mine inoperable.

Summoned by telephone, both Byron Pattison and Lemuel Shattuck went quickly to the mine and took charge of firefighting efforts. No one really knew where the fire was, and the manager and superintendent did not waste time in determining its location. They directed that three lines of two-inch fire hose be hung in the shaft, spray from which diminished smoke, lowered temperatures, and decreased velocity of air in the shaft. While water was poured into the works, Shattuck and Pattison studied plans of the workings to determine how to choke off the fire. They chose to bulkhead the mine at three points.

With arrival of miners and firefighting crews from other companies, work began on constructing three barriers to block air flow.

A bulkhead was driven at the 800-foot level where the mine connected with the Cuprite Mine of the Copper Queen. A second barrier was built at a sub-drift near the surface, and the mine collar was sealed. Within ninety minutes every opening to the surface had been tightly closed. Steam from the 250 horsepower boiler was then discharged into the shaft for five hours to assist in smothering the fire. No one knew how successful these efforts would be. Shattuck and his men could only wait and hope that lack of oxygen would extinguish the fire.

The next day a small hole was made in the bulkhead at the collar and a light cotton rope lowered to the bottom of the shaft to test for fire. It was retrieved unscorched. This was a time to be overly cautious, and Lemuel ordered that the mine remain sealed for another four days. When the shaft was uncovered, the smell of gas and smoke was strong but the odor cleared within three hours. Timbers in the shaft were sprayed with water, and firefighting crews entered the workings.

Shattuck and Pattison led the search for origin of the fire. Level by level the mine was combed. Finally, the point of combustion was found in the upper part of some lockers in the pumping station on the 800-foot level, probably caused by shorting of electrical conduits. Combustible waterproof covering of large cables spread the fire quickly to switch boxes in the station. From there the blaze traveled along wires to the 700-foot level, where it ate its way for 200 feet along electrical conduits in a drift. Had Shattuck and Pattison delayed efforts the night of November 19, the damage within the workings might have been far greater, or the mine lost altogether.[35]

Although the Shattuck-Arizona produced no ore during 1908 and 1909, forty men labored in its depths, repairing the fire damage and exploring for high-grade ore.[36] Since the first production of ore in 1906, only 7,937,782 pounds of copper had been extracted. During summer 1908 bodies of copper carbonate and oxide ore were discovered between the 600- and 700-foot levels. Beside having a seven percent copper content, this ore was high in iron, making it a prime smelting product. In August the southwest drift

This photo of Lemuel Shattuck was taken in 1910, about the time he was interviewed by Mary D. McFadden. Her words are a befitting caption. Lemuel Shattuck has "a large stock of magnetism and intensity set in an unobtrusive personality. He has the keen blue eyes of men from a country where space remains magnificent, and has all the natural habits of observation common to Indian fighters and cowboys, both of which he has been simultaneously. . . . Mr. Shattuck is above medium height and a typical western out-of-doorsman. Like the few men of his type, his voice is low pitched, but he seldom has to repeat his words, being sure of attention when and wherever he talks."   (Fathauer collection photo)

of the 600-foot level was driven through thirty feet of copper oxide. Large sulfide bodies were also discovered. By late May 1909 a half million tons of ore, ascertained by careful measurement, had been blocked out, and stood ready to mine — and only one-fortieth of the property had been explored.[37] By end of the depression the Shattuck-Arizona Mine stood in splendid shape, a virgin mine ready to resume production as soon as copper reached a satisfactory price.

The first sign that the Shattuck-Arizona was awakening from its slumber came early in December 1908 when a small tonnage of ore was shipped to the Copper Queen Smelter. Shortly thereafter Lemuel Shattuck announced the company was again hiring miners.[38] By end of the month the Shattuck work force had been increased by sixty men, and about seventy tons of ore were being shipped to the smelter daily. Copper prices remained low for yet another year, and the mine shipped only small quantities of its highest grade, self-flexing ore. At end of 1909 the company had produced 1,787,849 pounds of copper, valued at $232,072. There was no report of gold and silver recovery.[39]

The Denn Mine was even slower coming to life. Its pumps and boilers had to be cleaned and repaired, and water pumped from lower portions of the workings. By February that had been accomplished, and mining resumed on a limited scale on the 1200-foot level. The shaft was deepened by several hundred feet to facilitate access to several sulfide bodies encountered higher up, and to search for additional stringers of copper. For all of 1909 the Denn-Arizona produced a mere 99,222 pounds of fine copper and forty-one ounces of gold. In December 1910 the Denn's pumps were pulled and the mine again rendered idle.[40]

Closure of the mine was not due to lack of ore, or disappointment on the part of management. Quite the contrary. The Denn was considered a "splendid prospect," in that its claims covered an extension of what was believed to be one of the district's biggest ore zones. The hitch was that the zone was deeper at the Denn than at other mines, and ore extraction was impeded by heavy flow of water. Management hoped that deeper development at the Junction Mine would facilitate drainage of the Denn as well as prove the property.[41]

Although price of copper remained sluggish during 1910 and 1911, the Shattuck-Arizona steadily increased production and exploration. In 1910 the company added more than two miles of workings to tap high-grade ore lenses that increased reserves to nearly a million tons of oxide and primary sulfide ore. Directors of the Shattuck-Arizona, however, maintained a conservative policy toward extraction, forwarding to the smelter only ore ranging from sixteen to seventeen percent copper.[42]

Most production during this time was coming from a narrow seam of copper oxide. Because this exceptionally rich ore body was only fifty-five inches in width, little timbering was required, which earned the mine the distinction of producing the world's highest grade ore at the lowest cost. In 1910 it produced 4,751,000 pounds of copper at the incredibly low cost of 5.8 cents per pound. That year the Shattuck-Arizona paid its stockholders $350,000 in dividends.[43]

The time was ripe to take the Shattuck-Arizona off the curbs and list it on the Boston Stock Exchange, where the company would receive wider exposure. As previously mentioned, Shattuck and Merritt had gotten the ball rolling in that direction. After the dust had settled from the sale of Lemuel's stock, and with Merritt's aid, the company turned its wholehearted attention to drumming support in eastern financial circles. The broker was provided with financial statements, plans of surface and underground workings, estimates of ore reserves, and future earnings—everything he needed to sell the company to the board of governors of the Boston Exchange, the hub of national copper transaction. Meanwhile, from Bisbee Lemuel bombarded the mining world with such publicity as "Shattuck Mine is making copper for about four cents a pound, the record over any copper mine of the world today. The Shattuck is the cheapest copper stock on the market."[44] The activity of the Shattuck-Arizona Company—and its stock, which bounced between $27 and $40 a share—could not be overlooked in Boston. Merritt had no trouble in persuading the board of governors of the stock exchange to examine the company closely. In June 1910 Walter H. Weed and Frank H. Probert, New York City geological and mining engineers, traveled to Bisbee and inspected the Shattuck Mine for the Boston Stock Exchange.

Lemuel personally conducted the engineers through every foot of the mine. Their inspection was hampered by inexperience. Weed and Probert arrived in the Warren District expecting to see sharply defined ore bodies such as occur at Butte, Montana. While the Shattuck Mine had lots of rich ore, it was disseminated in irregular lenses and stringers. To Lemuel's dismay they complained that there was no "blocked out" ore. Shattuck explained that "ore bodies in Bisbee cannot be calculated as the ore in Butte and other camps, because we have irregular deposits." Weed and Probert were still perplexed. They could not calculate ore "in sight" from "irregular" ore bodies. Shattuck deftly solved the problem. As he related to Bardon, "I think they [Weed and Probert] can now see a little better than they could at first. . . . I told them I would donate each of them a little stock providing they reported the possibilities of the ore bodies not already blocked out to be what I thought they were. . . . I did not tell them what I thought the ore bodies extent was."[45]

A little "grease" did the trick. "The Shattuck property possesses real merit; future success assured," Weed and Probert reported to the board of governors of the Boston Stock Exchange. "The Shattuck Mine . . . is a solid, safe and substantial investment. . . . We commend the enterprise as a clean cut mining proposition with its future assured."[46] While the report was music to ears of Boston financiers, one more inspection was necessary. The Boston Exchange summoned mine management to an interview. Accordingly, Shattuck and company attorney J. M. O'Connell journeyed to Boston during the last week of June. They too passed inspection, for the Shattuck-Arizona Copper Company was listed on the Boston Stock Exchange on September 9, insuring a broader market for its securities.

To make Shattuck-Arizona stock attractive to that new market, the company launched a publicity program. In October 1910 President Thomas Bardon reported that "the property is equipped in first-class modern manner to handle 1,200 tons of ore per day whenever the market warrants. We have five miles of underground workings and over 800,000 tons of oxide and sulfide ores shown up, with a small area of our zone developed." In actuality, the publicity was aimed at selling enough stock to finance the erection of a smelter.

In 1907 plans had been drawn for a refinery of 700-ton daily capacity.[47] The company had even gone so far as to option a 251-acre site adjoining the Copper Queen and Calumet & Arizona refineries at Douglas.[48] The depression that year forced closure of the mine, and blocked awarding of contracts for construction. Building the smelter, however, had always been a bone of contention among Shattuck-Arizona directors. Shattuck and Byron Pattison, more than anyone else, knew the mine's richness and believed it capable of supporting a smelter. It was they who instigated the policy of conserving the mine's high grade ore. They also hoped that the Denn would eventually add to the flow of ore. Chisholm, Bardon, and other Duluth–St. Paul stockholders objected to the conservative policy of mine production. As manager of the mine, as well as its largest stockholder, Shattuck prevailed at the 1912 board meeting enough to prevent directors shelving the idea of a smelter altogether. The idea would be thoroughly investigated, construction bids requested, and a comparision made of costs of erecting a smelter versus outside refining.

Although there was discord over the merits of a smelter, announcement of plans to build a smelter could serve only to drive up the value of Shattuck-Arizona stock. Consequently the company beat its drums for two years, flooding the mining world with announcements of pending construction of the Shattuck-Arizona smelter. On June 12 it announced that a site had been selected in Douglas, south of the C&A smelter, "between it and the international line, . . . across the ravine [Whitewater Draw] from the Copper Queen smelter."[49] Cost of the refinery was projected at $400,000, and Arthur Houle appointed to superintend construction. When large bodies of lead ore were struck in the mine during summer 1912, a fifty-ton lead stack was added to the plans.

Announcement of the projected smelter, and the increase in price of copper to eighteen cents a pound increased Shattuck stock to twenty-eight dollars a share. When the company abandoned its plans to build a smelter in fall 1912, and contracted with the Calumet & Arizona smelter to process ore, not a ripple of discontent was heard from stockholders, for management had placed the company on a dividend basis of about five dollars a share per year. Payment of $175,000 in dividends in January 1913 drove Shattuck-Arizona stock above the thirty dollar mark. "Shattuck-Arizona has jumped from comparable oblivion into . . . one of the most active stocks on the local board . . ." noted the *Boston Financial News.*[50]

By mid-January 1913 the mine had 225 men employed, 190 underground. A stream of wealth poured forth at a daily rate of 250 tons of copper ore. Other metals were being added to the inventory: a silicious gold-silver ore had been struck on the 400-foot level and, 200 feet lower, layers of "sandy" cerussite showed up, assaying two dollars in gold, two ounces in silver, and eighteen percent lead per ton. A drift on the 300-foot level penetrated a great cavern — the largest ever discovered in the Warren District. Beneath this cavity lay a siliceous breccia zone containing shoots of high-grade copper and lead-silver ore, as well as a copper-lead vanadate, called cuprodescolizite. By end of 1913 an estimated 100,000 tons of this material had been blocked out and was awaiting shipment to El Paso.[51] The Shattuck-Arizona was on a roll. Within the next year, however, its momentum would be halted by events on the other side of the world.

Lemuel's saloon and gaming business had, by 1913, been feeling the impact of tighter moral strictures that had been evolving for some time. After 1905 the Warren District's mining companies had actively recruited married men with families: men who would live year-round in Bisbee; men who would help lay a permanent foundation for the community; men who would not dissipate their

Stalagmites of the Shattuck Cavern. A myriad of calcite forms, brilliantly colored blue and green by solutions carrying copper carbonate, occurred in this cavity, which was likened by one author to the "shadowy interior of a Gothic cathedral." (Fathauer collection photo)

wages on loose women and alcohol, but deposit their earnings in local banks or postal savings accounts. The disruptive vices of prostitution, gambling, and unrestrained liquor trade had to be controlled if Bisbee was to achieve the status of a mature community.

Incorporation of Bisbee and election of the city council sounded the first death rattle for prostitution. Through a series of ordinances and costly licensing procedures the redlight district was constricted and then isolated at the upper end of Brewery Gulch. Two other forces — the safety first and statehood movements — combined to change the moral tone of Bisbee and every other mining camp in Arizona.

Mining companies had long opposed saloon districts, not on moral grounds, but from the standpoint of economics. Saloons not only siphoned miners' wages, they were directly responsible for accidents and absenteeism. Payday meant drunken binges and failure to report to work. Invariably serious accidents and death occurred. It was not uncommon for hung over miners to fall down a shaft or chute, or to set off a premature dynamite blast. Such occurrences cut deeply into production, resulting in loss of revenue.

The mining companies attacked the tenderloins in any and every way possible: through the columns of their newspapers, by setting up a safety-first campaign, and establishing recreational activities such as baseball teams and clubhouses where employees could gather in an atmosphere free of alcohol. And the companies attacked saloon owners through political means such as lobbyists. The companies supported the drive for a higher standard of morality that was part of the statehood movement.

A liquor control act passed in 1907, requiring a uniform license of $300 a year for every saloon, was the first tangible result of the movements. That was followed by an anti-gambling law that forced every saloon owner in the territory to shut down that end of his business. On April 2, 1907, all open games in Bisbee ceased. Honest professional gamblers packed up and headed to other areas in the West — Nevada and New Mexico — where they could practice their skills; a few went to the border towns. Four saloons in town closed altogether. Lemuel complied with the law and shut down his games at the Shattuck, and closed his club upstairs. The third floor of the Shattuck-Schmid Building was converted to a lodging house.

As many of the dealers had been with him for years, Lemuel did not let them go, but gave them the choice of remaining in other positions. Joseph Martinelli, the Swiss head of the saloon's gambling concession, stayed on to manage the keg beer distributorship. E. A. Washburn, who dealt faro, declined a position and headed to San Luis, Mexico, where he acquired twenty acres of deeded land and established the San Luis Club, a combination saloon and gaming resort. He would later set up a distillery. The men that left employment at the Shattuck Saloon were all honest and hard working. Lemuel sadly admitted that an era had closed.

Besides losing his gambling concession, Lemuel had to contend with declining sales at both the saloon and beer distributorship. The economic slump of 1907–09 forced men from both Bisbee and Douglas, Shattuck's primary outlets. Out of work miners consumed little liquor. If that was not enough, Shattuck faced increased competition from other beer franchises penetrating his territory: Schlitz, Lamps, the Douglas Brewing Company, and Pabst. In June 1908 six saloons in the Warren District, each buying eight to ten barrels per day, went over to Pabst Beer. At $14.25 per keg, that was a considerable loss of revenue. The Phelps Dodge Mercantile Company, which had the Anheuser-Busch franchise for Cananea, made matters worse by shipping beer into Douglas, and undercutting Shattuck by $2.00 a keg. That summer he reported a fifty percent decline in sales to Anheuser-Busch.[52]

Days of the saloon were numbered. The perils of alcohol were being hammered home by such organizations as the Anti-Saloon League and the Women's Christian Temperance Union, as well as ministers and editors. When politicians got into the act, prohibition became a state and national issue. One by one states voted to ban consumption of alcohol. America was gradually drying up and, by 1909, the movement hit Cochise County when a registration for local option petition drew 2,000 signatures.

Most Bisbee saloon owners, including Shattuck, chose not to

beat their heads against a major political issue. Instead they followed the dictates of their brotherhood, the Knights of the Royal Arch. This organization, which included nearly every reputable retail liquor dealer in the country, was founded in San Francisco about 1904 for the express purpose of fighting politicians and ward heelers bent on making the saloon a political issue.[53]

On October 27, 1909, the Bisbee chapter of the Knights of the Royal Arch met at the Pythian Castle to draft a petition of rather startling magnitude.[54] Attempting to police itself, the organization suggested that the city council vote to close all saloons in town every night between midnight and five a.m. In that way Bisbee could enjoy a reputation of being a "tight" town. The city council acted on the petition, passing an ordinance on March 1, 1910, that closed all bars between midnight and six in the morning. The council, however, went a step further. It passed Ordinance 153, amending Ordinance 79, which had created the redlight district, and abolished the town's tenderloin, effective March 31.[55]

On April 1 the *Daily Review* announced the demise of the "Hottest Spot between El Paso and San Francisco." The obituary was tinged with sadness. "Last night marked the final appearance of Bisbee's tenderloin as a blazing myriad of mahogany, foaming liquor, and polished glasses. . . . No more dancing and drunken hilarity."

While closure of Bisbee's tenderloin was the most visible evidence of the work of the Knights of the Royal Arch, the organization projected its influence far beyond the Warren District. Through a levy of fifty cents per keg upon retailers, and twenty-five cents per keg on wholesalers, the Knights raised a war chest to influence politicians and elections. Shattuck and J. J. Walsh were empowered by the group to collect this tax in the Warren District.[56]

Despite lobbying efforts by the Knights of the Royal Arch, prohibitionists prevailed on the fall 1914 ballot, and state-declared prohibition became reality at midnight December 31, 1914, as bar owners herded their quiet customers out the doors, locked up, and went home. "Drunken hilarity" ceased everywhere in Arizona.

Just how Lemuel Shattuck felt about loss of the liquor business is difficult to assess. He was, after all, a man who kept his feelings to himself. Next to the lumberyard, the liquor business had figured importantly in building the Shattuck fortune. Lemuel had sold over a million dollars worth of keg beer alone through his Anheuser-Busch distributorship.[57] The saloon was a man's world, and Lemuel was very much a man. His correspondence reveal a man at home in his liquor business, a perfectly respectable enterprise in the early years of the 20th Century. To collect empty and full cooperage, pull ice boxes and tapping outfits from longstanding clients, and return them to Anheuser-Busch, must have been a bitter pill for Shattuck to swallow. Undoubtedly he experienced a profound sense of loss when he moved his prize liquor stock to the barn behind his house and padlocked the door of that structure.

Lemuel Shattuck was not the type of man to convert his saloon to a soft drink emporium, as many saloonists did. Even before advent of prohibition, he had begun casting about for other enterprises. The lumber business had changed too. When Lemuel first wandered into Bisbee there was one lumber company. Then the chief materials of the building trade were adobe, and lathe and battan. By 1907 half a dozen companies were locked in fierce competition, selling not only lumber, but brick, pressed concrete block, and cement. A series of fires at Bisbee, and soaring insurance rates, had hammered home the wisdom of using fire-resistant building materials. The depression of 1907 not only put an end to plans for a lumber enterprise in the Sierra Madre, it created a decline in sales at both the Shattuck and the Douglas lumber companies — reasons enough to contemplate getting out of that business altogether. In 1909 he sold the lumber business on upper Brewery Gulch to Jim Henderson, who renamed it the Pioneer Lumber Company. In March the following year Henderson combined the company with that of E. A. Watkins. Thereafter it was known as the Henderson-Watkins Lumber Company.[58]

Shattuck jumped out of one enterprise, only to plunge into another. Reflecting on the cowboy days of his youth, Lemuel often yearned to have a spread of his own. The time had not always been right for his move into ranching, however. The 1890s were hard years, during which overgrazing destroyed grasslands and drought decimated herds. Both had caused the Erie Cattle Company to depart the Arizona scene by 1901.[59] Opening of new markets and the introduction of improved breeds of cattle benefited the cattle industry in the late 1890s. The price of beef steadily increased, until mature steers were bringing an average of $35 per head, and older stock up to $50. By 1908 Arizona ranchers were prospering.

As a banker, Shattuck had aided many cattlemen in meeting the heavy capital investments involved in new methods of cattle raising, and had watched with keen interest this transition from famine to feast. Prosperity of Arizona's cattle industry spurred Shattuck to enter the cattle business in 1908, not as a rancher, but as a broker.

In May of that year Shattuck was buying cattle from ranchers in the Sulphur Springs Valley on behalf of ex-Arizona Ranger Captain Burton Mossman, who was affiliated with the Hansford Land and Cattle Company of Lakewood, New Mexico, and Evarts, South Dakota. In October 1908 Shattuck purchased cattle for the McPherson & Hysham Company of Omaha, Nebraska.[60] Shattuck's brokering days were shortlived, however, for in 1911 Lemuel became a full-fledged rancher in partnership with William Lutley and John Meadows, men he had known since days of the Erie Cattle Company.

Englishman William Lutley was ten years older than Shattuck. He had emigrated to the United States in 1877, and worked as a lumberjack in the redwoods of California before coming to Tombstone in 1881. He mined and freighted before joining the Erie Cattle Company. In 1890 he met John Meadows, a quiet but gregarious man, who had grown up in Texas cattle country. As a youth, Meadows participated in trail drives to Abilene and worked for large cattle spreads, including the Sawyer Cattle Company, at Barnhart, Texas. Shattuck, Lutley, and Meadows were a sagacious trio.

In March 1911 they acquired the Bar Boot Cattle Company, located in Box Canyon between the Swisshelm and Chiricahua Mountains, northeast of Douglas. History of ownership of this ranch is somewhat clouded. Lutley may have had an interest in the spread in the early 1890s, for it is reported that on one of his early freighting trips through the area, he located the Box Canyon Ranch, and with Clint Stewart, managed the Hi Lonesome outfit five miles away on the west side of the Chiricahuas. About the same time he was reported associated with other "pioneer cattlemen" (perhaps Erie principals) in the Swisshelm Cattle Company. In 1908 the Bar Boot brand was registered to C. L. Cummings & Company of Tombstone. Lutley may very well have been part of that venture.

At any rate, Shattuck money enabled the trio to purchase the entire holdings of the ranch. From time to time the three partners added other smaller allotments and tracts to fill out irregular boundaries. In all, fourteen smaller outfits were added to the original acreage. Among the latter was the O'Keefe Ranch at the entrance of Rucker Canyon, purchased by Shattuck and his associates in 1914 from Joseph Curtner.[61] The Bar Boot and the O.K. Ranches were run separately, the latter spread under the direction of John Meadows. Together, the ranches held title to 2,400 acres and leased 20,000 acres of grazing land from the Forest Service.

Open ranges and large herds were a scenario of the past. Bar Boot acreage was fenced with barbed wire, and stocked with Herefords, a breed that adapted well to Arizona's environment and climate. The herd grew steadily, reaching a maximum size of about 2,400 head. Cattle sold from both ranches, mainly as stockers for northwestern growers, were driven into Douglas by mounted cowboys, who camped overnight on the way. Although occupied with mining and banking matters, Lemuel always rode roundups in late October or early November and attended most of the cattle shipments.

Shattuck's entrance into cattle raising may have been motivated by more than business concerns. After all he was a farmer's son, who may have experienced a sense of loss from closure of his liquor business. The second reason is more complex. He may have felt a desire to pull together some loose ends in his life. The few years spent with his half-brothers at the Erie Cattle Company held special meaning for Lemuel Shattuck throughout his entire life. On the open range of Sulphur Springs Valley he had served his apprenticeship in the cattle trade.

Henry (right) and Warner (left) shyly pose on a stack of lumber in the shed at the family lumberyard about 1896.   (Fathauer collection Photo)

SEVEN

"PAPA"

IT WOULD be easy to revert to the cliche that behind every successful man stands a woman. Perhaps? Lemuel nevertheless possessed several basics for success: an innate intelligence and a philosophy conditioned by Universalist concepts — he accepted people. The necessity of dealing with all types of men sharpened his ability to measure a person's worth. Lemuel's acumen was honed on the frontier. He could sense what was needed, and stepped forward to provide it. Facing hazards of range life, the saloon business, and mining pushed trepidation from him. These attributes probably enabled him to choose a wife who meshed with him in more than just a physical sense.

Frugal, unpretentious, and well educated for her day, Isabella Grenfell Shattuck was imbued with a work ethic similar to Lemuel's. She adored her husband and children, and was protective of both. Although she objected to the liquor and gambling business, she assisted her husband in other family enterprises. She managed rental properties, collected rents, and supervised repairs. Finishing her early morning household duties, she went daily to the lumberyard to look after the running of the business. Drawing heavily on mathematics from her teaching background, she measured board feet of lumber as easily as a seamstress would cloth. Isabella took orders, delegated work to employees, kept the books, and attended to billing.

There is little doubt that by 1907 Isabella Shattuck had her own juggling act in motion. She had her hands full with teenage sons Henry and Warner, as well as nine-year-old Mark. Intelligent and precocious, the two older boys scrapped between themselves and with other children in town. They did not always behave at school. She controlled fisticuffs at home with the stern warning, "Wait until your father comes home." A special look in her eyes told the boys they better shape up.

Her task was somewhat simplified by the move of her mother and father to San Diego in 1904. Lemuel purchased his father-in-law's seven-room, wood-frame house down the Gulch and moved the family in. It was a cheerful home with tongue-in-groove ceilings. Each room, except the kitchen and bathroom, were wall papered. Large throw rugs covered the floors; the living room was graced with a black leather sofa and an oak cabinet full of mineral specimens. The home was one of the first in Bisbee to have a bathroom with running water, a great footed bathtub, and a flush toilet. There was electricity and a telephone.

Since Isabella loved gardening, Lemuel leveled areas in front and back of the house. Topsoil was brought in and spread out, and Isabella put in a small garden and laid out walkways interspersed with flower beds, all overshadowed by a large tree. The backyard was fenced in and spread with sand for a children's play yard.

In August 1904 a blond, blue-eyed son was born. He was named Carl. Scarlet fever, however, claimed his life eighteen months later. As with earlier deaths in the family, Lemuel recovered quickly, but Isabella remained despondent for some time. Each death in the family left her a little more sad and protective of those close to her. She found solace in gardening and in caring for pets. Belle bought plants, and often sent to California for pedigreed puppies and kittens. Her grief was lifted on October 7, 1906, with the birth of Spencer.

With four sons, Lemuel was convinced that he was destined never to have a daughter. His wish would be granted, however. In March 1909 a daughter was born, an event celebrated by both father and the town. On March 23 the *Daily Review* proclaimed in its social section that "L. C. Shattuck is the proudest man in Bisbee. A bright, healthy, girl baby is the source of his happiness. Little Miss Shattuck arrived Sunday night. Mr. Shattuck is the father of four boys, and it has been his heart's desire that a girl baby should be added to his family. Both father and mother are overjoyed over their good fortune. . . ."

The house overflowed with well-wishers: miners, merchants, cattlemen, and prospectors. All inquired as to the infant's name, but both mother and father were at a loss as to what to say. Since the early days of pregnancy it had been assumed the baby would be a boy. Caught by surprise, the parents had no name.

Lemuel and Isabella wanted a family name. Phoebe, Lemuel's mother's name, was suggested, but Isabella objected; and Lemuel didn't like Isabella's mother's name Elizabeth, or its shortened form, Eliza. No family name seemed to fit the child. Consequently, the baby was registered in Tombstone as "Baby Girl Shattuck." Until old enough to be enrolled in kindergarten, she was called "Sister" by her parents and brothers. At that time, mother sat down with her child, and they decided she would have her mother's name, shortened to Isabel.

A second girl was born on April 19, 1911. Except for an attending physician and midwife, the house was vacant, for Lemuel and other Bisbeeites were at Douglas observing the battle at Agua Prieta between Mexican federal forces and revolutionaries under Red Lopez.[1] This time there was no problem in naming the baby. Called Dorothy, she was the last of the Shattuck children.

Because Brewery Gulch bore the reputation as being the "Hottest spot between El Paso and San Francisco," some Bisbeeites — especially supervisory personnel of the Copper Queen and C&A mines who were building palatial homes in Warren — looked with disdain on its residents. As a prominent businessman, Lemuel Shattuck could have moved his family to Warren. For Isabella's sake he did not. She may have objected to hardships of mining camps, but she loved the Grenfell home, which was close to Shattuck businesses planted along the Gulch, and family friends such as the Muheims, Carretos, and the Pritchards. Although the spacious house, with a veranda in front and a porch in back, sat between the saloon and redlight districts, Lemuel and Isabella felt no shame in living on the thoroughfare. And their lifestyle reflected that pride.

The relationship between Lemuel and Isabella was especially close. As mates and working partners, they discussed everything from current news to decisions regarding banking and mining; saloon matters were better left unsaid, however. Isabella and Lemuel played whist and read to one another from the best magazines and newspapers of the day. Although the Shattucks had a sizable library, she frequently borrowed books from the Copper Queen Library. They always found time to help with their children's homework, and instilled in them the same ethics that had guided their lives and the lives of their forebears.

Although Isabella and Lemuel were products of strict religious households—she of Cornish Methodist background, he of Quaker-Universalist—both felt regular church attendance did not make the Christian. Daily ethical practices instead determined an individual's merit. Good moral conduct was instilled in Shattuck children from an early age. When five-year-old Warner lifted an apple from Hickey's grocery store, Isabella took him back to the store. Handing back the apple, Warner tearfully apologized to the proprietor. Contrasting with the strict Shattuck code of conduct was an odd informality. Henry and Warner called their parents by their first names—Belle and Lem. A similar practice was carried on in the Grenfell household, both families accepting it.

Lemuel and Belle expected their children to excel. Teachers were to be respected and obeyed. If the Shattuck children committed a misdemeanor in school and were punished, they likewise were reprimanded at home. When the two eldest sons were expelled for hanging the principal head first out the school window, Lemuel saw no humor in the prank and came down hard upon Warner and Henry, denying them privileges. The boys were reinstated in school, toed the line, and graduated with honors. Belle was never informed of the incident.

Like their parents, the Shattuck children were raised to be independent yet not aloof. Perhaps that was the reason both parents encouraged their children to walk to and from school and to mingle freely with children their age. The names of their playmates and friends reflected that decision: Avilez, Lopez, Ojeda, Medigovich, Koch, Carreto, Sugich, Kritchlow, and MacDonald.

Because the border was nine miles away, and Spanish spoken as much as English in the Warren District, Lemuel insisted that his children learn Spanish, which was taught in junior high and high school, although their mother preferred they study Latin to gain a greater understanding of their language and literature. The children were educated in Bisbee's public schools, and also sent to the Sisters of Loretto school near the Catholic Church on Quality Hill, where they received music lessons. Isabel went to Central High School; Dorothy was enrolled in Miss Riley's School for Young Ladies on upper Main Street. All Shattuck children were expected to go on to college or universities. That was not an unusual expectation; in those days higher education was revered, and many hardrock miners scrimped and saved to send their children to institutions of higher learning.

Protective is the best word to describe Isabella Shattuck — protective not only of her husband and family, but of others. It is a well known fact in Bisbee, even to this day, that Isabella turned no one away from her door who was in need. She always had food and money for hoboes who appeared at the back gate. They arrived in such numbers that Lemuel often said that there must be some special mark on the kitchen door that encouraged them to come to the Shattuck house. Isabella also kept a drawer full of new children's shoes and stockings, which she gave to barefoot Hispanic children passing her house during winter. When the children reappeared barefoot several days later, Lemuel explained the children's parents had probably sold her well-intentioned gifts. She could not comprehend parents committing such a thoughtless act against their offspring. She nevertheless continued dispensing shoes and stockings to needy children on cold winter mornings.

Yet another story exemplifies Isabella's concern. One afternoon while at the lumberyard, she saw two drunken miners walking up the Gulch. One leaned over and picked up a wandering gila monster, which immediately bit him on a finger and hung on. Try as he might, the other miner could not pry the creature off. The man whom the lizard bit urged his companion to pick up an axe embedded in a nearby tree stump and chop the creature's head off. When the man grabbed the axe, Isabella rushed down the path, fearing their lack of sobriety would result in the victim's hand being cut off. She and another passing miner finally pried the gila monster from the man's finger.

Isabella's protectiveness cloaked an innate fear that was manifested in peculiar ways. Every summer Indians arrived in Bisbee with buckboard wagons filled with ollas that were purchased by Bisbeeites. The ceramic jars were tightly wrapped with burlap, filled with water, and hung on verandas. Recalling that old George Warren, discoverer of ore in the Warren District, had been an Indian captive in his boyhood until ransomed for a sack of sugar, Isabella always pulled her children close to her when Indians appeared. Regardless of Lemuel's reassurance that these Indians were interested only in selling their wares, she remained suspicious.

Overly protective women can be demanding and weak. That was not the case with Isabella. She pulled her weight and did everything expected of her. She attempted to shelter her daughters from mining town vices. Once when mother and daughters sat beside the window, a madam in all her finery walked past the house. "What a pretty lady," Isabel exclaimed. Mother merely frowned and said nothing. When Lemuel arrived home, she told him their eldest daughter thought a madam was a pretty woman, and left the situation in her husband's hands. Without undo fanfare, he identified the woman and her occupation, and explained she, and others like her, were necessary in mining camps where majority of the population were young, single males.

Belle was fastidious in her appearance, that of her children, and of her household. The Shattuck home was always neat and clean. Rugs were taken into the backyard several times a year and beaten free of dust. The children's clothes were always immaculate and pressed. Each morning they were given clean linen handkerchiefs as they went out the door. At first Isabel went off to Central School wearing a white dress, and embroidered underpants trimmed in lace. When she indulged in gymnastics at recess, swinging on the rings and hanging by her knees from the bars, she was sent home with a note requesting she be dressed in bloomers. To Belle's horror her eldest daughter competed with her brothers, played baseball, touch football, and climbed over the school yard gate as did the boys. Isabel never quite developed into the "proper woman" her mother desired.

Dorothy on the other hand, satisfied Mother. She was mannerly, spoke perfect English, practiced piano one hour daily before school. She had a beautifully furnished two room playhouse, where she played with dolls and imitated Mother. Every Saturday morning Isabel had to join Dorothy in the play house. Perhaps Belle thought some of Dorothy's femininity would rub off on her older sister.

In contrast to Isabella's protectiveness, Lemuel was a pillar of strength. He possessed uncanny insight into his children's character: he knew their strengths, weaknesses, and foresaw potential follies. Unlike Belle, he forgave and forgot misdoings quickly. During times of family crisis, Lemuel always supported Belle. When Isabel developed a severe respiratory illness, and Nelson Bledsoe and another doctor attended her all night, fearing the child would die, Lemuel was there, strong and stalwart, not showing his inner feelings. He was always there for Belle and his children.

As Mother endeavored to instill femininity in her daughters, Lemuel promoted his sons' masculinity. As soon as they were old enough, he took them on tours of the mines, where they talked of geology, minerals, copper, and business. Mining soon permeated the boys' thinking. Observing Warner digging up the garden, Lemuel commented that his son had the makings of a geologist. When the boys asked about saloons, gambling, and the painted ladies, Lemuel withheld nothing. Mining camps were a man's world they should know about.

Father and sons never missed major social events of the year, such as Fourth of July, which was always ushered in with the echo

Isabel and Spencer show off their prize bull at the IV Ranch some time in 1946. (Fathauer collection)

Dorothy Shattuck, age seven. (Fathauer collection)

through the canyons of pistol shots and the firing of anvils. They donned their best clothes and hurried downtown to attend the long parade through Main Street. With single- and double-jack drill contests, tugs-of-war, the Fourth of July was an all-day affair, capped off with a masked ball at the Pythian Castle, followed by fireworks. Just as avidly followed was the yearly circle of baseball games between teams fielded by the mining companies. In the contest between the Blue Ribbons and the Giants, the boys rooted when their father came to bat. So subtly did Lemuel build his sons' interests that there was seldom friction between them.

Lemuel was both father and mentor, and had little trouble in guiding and counseling his sons. Eldest son Henry was to be a farmer like his ancestors, and he studied agriculture at the University of Arizona. Geology fascinated Warner and he was enrolled in mining at the University of Minnesota. Mark and Spencer majored in business and economics: Mark at the University of Chicago, Spencer at Stanford.

Isabella and Lemuel were of the opinion that daughters should be as free-thinking as sons, able to form their own opinions so they could stand on their own feet as adults. There would be no "pitiful Pearls" in the Shattuck family. Both parents gave of their intellect as much to the daughters as to their sons. Daughters, more than sons, have a special capacity to tap a father's softness. And such was the case in Lemuel's family. To Isabel and Dorothy he was "Papa."

While Dorothy was her mother's shadow, Isabel was Lemuel's. Every day he walked with Isabel to her first grade class. Doors and windows along Brewery Gulch opened, and people called their greetings: "Good morning, Mr. Shattuck," "Good morning, Lem." At the lower end of the Gulch they stopped and gazed in the window of the Vienna Bakery. Neither the fresh Napoleons nor Mexican dulce rolls tempted Isabel as much as the big sugar cookies, which her father frequently bought. There they parted. He continued down the Gulch and up Main Street to his office at the Miners and Merchants Bank. Isabel climbed the steep steps to

Opera Drive and Central School, which housed kindergarten through high school classes for all Bisbee's children.

Isabel adored her father and, if permitted, followed him everywhere about town: to the bank, and to the mine. She accompanied him to bank directors' meetings. At the end of a meeting on one occasion, the men voted to pay Isabel the same stipend they received. Lemuel declared that this was carrying things too far—Isabel was there to listen and learn. Trips to the Shattuck-Arizona Mine, high in a canyon on Bucky O'Neill Hill, were the highlights of their relationship.

Because miners believed it unlucky for women to enter underground workings, Lemuel never took his daughters below the surface. The above-ground complex held fascinations enough, especially the assay office with its array of apparatus and colorful, sometimes delicate minerals. Pungent smells of chemicals permeated the office. Lemuel was like a child in a toy shop when it came to the Allis-Chalmers steam engine and its great doubledrum, with hundreds of feet of steel cable that propelled the cages in the warm, steamy shaft. Although Isabel did not quite understand the intricacies of the monstrous plant, his enthusiasm infected her nevertheless.

A growing family rendered the Shattuck home anything but dull; and the activity was intensified by comings and goings of friends, business associates, miners, and prospectors. Many of these men were colorful characters with unique backgrounds whose visits left lasting impressions on the Shattuck children.

Dutch Koch, a colorful old sourdough with a glass eye, was one such person. He wandered in and out of Mexico, always in search of a bonanza. In 1908 he made what he thought was a promising discovery in a remote area of the Sierra Madre in Sonora, and hired a few peons to work his claim. Revolutionary uprisings waxed and waned, each upheaval leaving his workers more petulant and lazy.

Dutch would overcome their avoidance of work with his glass eye. Taking it out, he placed it on a tree stump and told the peons that it was a magical eye that watched them even if he was not there. Terrified of such a wondrous object, the men became more amiable and hard-working. Unfortunately richness of the property was only surface deep, and when the revolution heated up Dutch returned to Bisbee.

He wandered in and out of Lemuel's office at the bank, as did many other prospectors who never struck it rich. More often he appeared at the Shattuck house, where he spent hours spinning yarns. His special rhythmical knock at the door was instantly recognized by the children, who gathered to hear his latest adventure. Sometimes he would take out his glass eye to polish it, at which time he allowed the youngsters to inspect it. Isabella disapproved of this, but never protested very loud, for she was as eager as her offspring to hear his tales.

Dutch took off again, this time to Central America, and it was more than a year before his rhythmical knock was heard at the door. He was thin and sallow looking, hardly recognizable. Spellbinding as ever, he related how he and a partner had set up housekeeping in a one-room shack in the jungle. It rained continuously, making prospecting difficult. Armies of huge ants ate their way through furniture and everything made of wood. Dutch and his partner placed the table legs in large tins of water and climbed up on the table until the horde moved on. Torrential rains made paths and roads impassable. The men ran out of food and other supplies. Desperate, they killed red monkeys for meat. It did not make any difference how they cooked it, or for how long a time, the meat remained tough and stringy. Dutch's teeth hurt and became so painful that he persuaded his partner to pull them all out with a pair of pliers. As soon as weather permitted, he headed back to Bisbee, toothless but undaunted.

Mr. Glitz, a German, was another venturesome miner that frequently called on the Shattucks. Deciding to prospect in Alaska, he withdrew his savings from the Miners and Merchants Bank, bought lumber and built himself a small boat, which he invited Lemuel to examine. Shattuck did so, and at the dinner table that night he described Glitz's plan for a solo voyage to Alaska.

Isabella was astounded and asked her husband if he had attempted to discourage the man from such a perilous enterprise.

Lemuel insisted that the boat was well constructed and he was sure that Glitz would make it. He proved to be right. Several months later a letter arrived from Alaska addressed to the Shattucks. Souvenirs followed: napkin rings made of walrus tusks, and exquisitely carved figures of seals, fowl, and bears.

Another frequent visitor of the Shattuck household was Leo Jeleniewski, a handsome patrician with gleaming white hair and a well-trimmed mustache. Not as adventurous as some miners and prospectors, this Polish immigrant managed Lemuel's wholesale liquor business and worked part-time at the lumberyard.[2] He had a deep affection for the Shattucks, and always showed up with presents for the children. The man's courtly manner endeared him to Belle and the children, but they knew little about his background, for Leo revealed nothing of his past. It was learned later that he had sent considerable money to relatives in Europe.[3] Unable to pronounce his name, Dorothy always referred to Leo as "Mr. Gallon of Whiskey." He never took offense and continued to call at the house bearing gifts.

Lemuel projected his sense of guidance beyond Belle and his children, to the Keatings and Grenfells, his relatives by marriage. John Keating, Lemuel's business partner, brother-in-law, and close friend, experienced a rocky marriage to one of Belle's younger sisters, which finally culminated in the woman abandoning her two daughters and departing Bisbee with another man. With help from Belle, Lemuel, and other Grenfells, Keating cared for the children as best he could. Upon Keating's death in summer of 1903, administration of his estate fell to Shattuck.

As a banker, Shattuck administered many estates. He was privy to inside information regarding the growth and moves of major mining and industrial companies, which allowed for investment of capital in stocks yielding maximum dividends. Ties with other banking institutions provided investments in high yield bonds and interest-bearing notes. Above all Lemuel was scrupulous. His sense of friendship and family loyalty took Shattuck beyond mere administration. Instead of returning Mary and Margaret to their mother's custody, Lemuel assumed guardianship of both girls.

Irishman John Keating endeavored to raise the girls as Catholics, and upon his deathbed, requested Lemuel use the modest Keating estate to remove eight-year-old Margaret and seven-year-old Mary from the unwholesome atmosphere of a mining town, and assure their education in parochial schools. Lemuel fulfilled his friend's wish by enrolling the children in 1907 at St. Mary's Academy in Los Angeles, and providing funds for their tuition, board and room, and personal expenses.

Shattuck's correspondence with Sister Catherine, Mother Superior of the school, suggests he acted more as a father than a guardian. When Los Angeles weather turned cold Lemuel sent money to buy the girls warm clothes and coats. He frequently provided small sums for entertainment, and for those frivolous requests every daughter makes of her father. Lemuel constantly monitored the girls' progress in school.

"Glad to hear the children are well and that they are getting down to their school work," he wrote Sister Catherine. "Have them write me a letter once in a while, as it will teach them correspondence."[4] Correspond they did. "Dear Uncle" letters arrived at least twice a month, and Lemuel reciprocated by commenting on their progress in school. He kept them informed about events in Bisbee, and activities of Shattuck and Grenfell children. His letters to Margaret and Mary, in many respects, convey a softness not present in correspondence with his own children. In situations threatening the girls' welfare Shattuck was steel hard.

When the girls' mother appeared at St. Mary's Academy desiring her daughters vacation with her, Shattuck emphatically told Sister Catherine: "I do not want the children to spend their vacation with their mother, nor do I wish you to even let them go down town with her, as I do not consider their mother proper company for the children."[5]

On attainment of high school age, Shattuck transferred the Keating girls to the Academy of Our Lady of Peace in San Diego.

The move broke the loneliness and isolation of Los Angeles. They were now close to their Grenfell relatives, and could attend weekend family outings, picnics, and beach parties, something they had not experienced since leaving Bisbee.

Lemuel guided Margaret and Mary through their childhood with few difficulties. Even after they attained adulthood — Margaret turned twenty-one in 1916, Mary a year later — he continued to advise them on personel and financial matters. Both girls respected and followed his counsel long after they were married. That would not be the case with other relatives.

Lemuel's father-in-law, Edward Grenfell, died on August 7, 1904, leaving an estate consisting of San Diego property, mining stocks, and considerable cash to his wife and six unmarried children. In 1908 Edward's wife, Elizabeth, died. The following year oldest son Albert (Bert) Grenfell drank himself to death, leaving only William and Margaret, affectionately called Daisy, to look after Clarence, Russell, and Rufus. Will was a hardrock miner and a drifter like his father and could not assume guardianship of his young brothers. Consequently, care of the three boys fell to Daisy, who was also appointed executor of the Grenfell estate.

Apportionment of the estate: one-sixth to Daisy; a sixth to Clarence, a third to Rufus, and a third to Russell, drew in Lemuel as financial adviser to the Grenfells. The estate could not be closed, however, without appointment of a legal guardian for the three minors. Shattuck was roped into managing a third family when the boys unanimously voiced his name to the court as their choice of guardian. On October 29, 1909, the estate was closed, and two weeks later Shattuck assumed guardianship of the three boys, who would reside with Daisy in San Diego.[6]

Guardianship of the Grenfells was far different than looking after two girls in religious institutions. Rufus, the youngest boy, was just entering high school and did not present immediate problems, but older brothers Clarence and Russell were running with a "bad crowd." They stayed out late at night, spending most of their time and money at Tent City, the amusement section of San Diego. A continual fight raged between them and Daisy over the estate, and the sale of several family properties which netted $5,393. Each wanted a larger share, and advances against their inheritance, all of which were impossible to grant until conclusion of probate. By summer of 1910 Daisy was at wits-end and seriously considered turning over to the boys her share of the inheritance, which amounted to over $45,000.

As financial adviser, guardian, and friend, Lemuel quickly moved to prevent Daisy from making the biggest mistake of her life. He tersely drove her into line:

You are not going to turn over your share of the estate to the small boys, nor are you going to divide the $5,393.30 among the . . . boys if I can help it. You are surely entitled to your share of the estate according to the will, and you are very foolish to even think about turning over what belongs to you. It is a good deal better to keep what is yours, and if the boys should ever go broke you can then help them. You are better able to take care of it than the boys, and besides you are a girl spending the best years of your life for the boys, and the boys should be better able to rustle for themselves than a girl. . . .

Now don't be foolish, Daisy. Keep what is yours as you do not know when you might need it, and if you do, you will need it bad. If you should get married you will need all your money to take care of your children. If you do not get married, you will need it to keep yourself in your old age. Take my advise and keep what is yours, and then if you should see that any of your brothers were in need you can help them out. . . .

I will not have it any other way. You have got to do as I say because it is right, and if I ever hear that you are again so foolish as to ever think or mention it again, I will come down to San Diego and have a first class fight with you and knock the stuffings out of you in the bargain.[7]

Although Shattuck's stern letter saved Daisy's inheritance, it did not solve the Grenfell problems. The boys bickered among themselves, and together argued with Daisy over money. With Lemuel's guidance, she furnished allowances, and additional money when needs were justified. Clarence, however, was ungovernable as a teenager. When not working as an apprentice plumber, he caroused, and upon attainment of adulthood squandered his small inheritance on a ranch and a Packard Auto Rental Company, both of which failed.

Lemuel attempted to guide Russell away from the path taken by Clarence, but without much success. Russell also had difficulty managing money, business, and his personal life. He drank too much. His San Diego Motorcycle Company, organized with Grenfell money, collapsed, and so did his marriage.

The failures of Clarence and Russell were nothing compared to those of Rufus. In grade school he presented no problems. He was studious, almost angelic. After his sixteenth birthday, he turned into a wilful, spiteful individual, utterly contemptuous of adult authority. In fall of 1914 Lemuel attempted to remove him from the influence of his older brothers by proposing Rufus be enrolled in a military school, two of which were selected. The Army and Navy Academy at Pacific Beach was close to the Grenfell home, but the Shattuck School in Minnesota was Lemuel's preference. If sent there, Rufus could be supervised by Warner who had just enrolled at the University of Minnesota, fifty miles away. Rufus went to the Shattuck School but stayed there only a few days, attending no classes. Returning to San Diego, he defiantly informed everyone "that he would not go to school; that he was willing to work," and if his guardian forced the issue, "he would not study." Furthermore, he did not care what Shattuck did to him.[8]

Rejecting all adult supervision, Rufus masked his thoughts with a cherubic smile. He was withdrawn. "Youthful indiscretions" were thought to be the cause of his bashful, timid personality. At first Rufus was able to bluff family attorney James Wadham, who remarked, "there was something pathetic about this boy that I do not quite understand." The lawyer suggested that some adult have a "good, quiet, wholesome talk" with the youth. Lemuel Shattuck was capable of frankly discussing any subject, but was unable to lure Rufus into a situation where both could sit and discuss what was really bothering the boy. Rufus's conduct became more outrageous, finally driving Wadham to admit that "Rufus does not know what the word obey means. He has never been obliged to obey anyone; he has never been under the jurisdiction of anyone that would deal with him from a maternal or paternal standpoint."[9]

Throughout 1915 Rufus sponged off Daisy, and Shattuck furnished him money from the Grenfell inheritance to purchase a motorcycle to commute to work at a automobile repair shop. There was little work provided, however, and he spent a great deal of time with a teenage girl, who became pregnant. In June 1916 nineteen-year-old Rufus swore before a San Diego County clerk that he was twenty-one, obtained a marriage license, and the couple eloped. The girl's mother contacted a lawyer and threatened to jail Rufus.

After spurning all advise and guidance, Rufus expected his guardian to bail him out of this jam. "I think she is going to sue me because she thinks I have money," he wrote Shattuck.[10] It took gall to request his guardian to determine "her intentions."

Shattuck bristled over this mess. "The fact of the matter is that you are married," he wrote Rufus. "I believe it is also a fact that you can be prosecuted for perjury as you swore before the clerk . . . that you were twenty-one years of age. If I am not badly mistaken, I as your guardian could bring suit in the courts and have the marriage annulled because . . . you are under age and never obtained my consent to the marriage."[11]

Before Shattuck could move to have the marriage annulled, the girl's mother filed suit for $5,000 in damages. Rather than fight the case, which would draw adverse publicity, Shattuck and attorney Charles Pritchard hammered out a settlement of $4,160. When Rufus proclaimed that he would never live with the girl "under any consideration," and the girl stated she would have nothing to do with Rufus, a divorce complaint was filed, and Lemuel forwarded a check to cover the settlement and attorney fees. An interlocutory decree of divorce was rendered on October 28, 1916.[12]

Lemuel made one last attempt to remove Rufus from what he considered bad influences. He invited his nephew to Bisbee, and in all probability would have provided him a job. Rufus, however, had one more rebuff saved for adult authority. On December 14, 1916, he informed Lemuel that he was not coming to Bisbee "because my wife is here and I still like her and am going to live with her the first chance I get." This was the proverbial straw that broke

the camel's back. As hard as Shattuck could be, Rufus had penetrated his armor. Lemuel was both mad and hurt, and felt he had
let Rufus down. On December 28, he informed James Wadham
that "I have been a failure as far as his [Rufus's] guardianship is
concerned and feel that it is my duty to turn the guardianship over
to the court who may appoint a guardian for him."[13]

The bruising received from the Grenfells in no way dampened
Lemuel's concern for them. When Will was unemployed, Shattuck
hired him as a foreman for the Denn Mine. Clarence was employed
both as a mechanic and a guard at the Shattuck Mine. As a member
of the Citizens' Protective League, he participated in the deportation of Wobblies from Bisbee in 1917. Daisy remained in close contact with Lemuel, and eventually moved to Tucson. Rufus kept his
distance, however. While Russell at first failed in business, he
loved automobiles, and with Shattuck's help seized an opportunity
prior to World War I to acquire a Ford automobile franchise in
Western Mexico. His agency in Culiacan prospered, protected by
both federal and revolutionary leaders. Russell married a Hispanic
girl and raised a large family, the descendants of which still live in
Mexico. And he introduced baseball to the region.

The years prior to World War I were eventful, satisfying years
for Lemuel Shattuck. His desire for a daughter was fulfilled not
once, but twice. Henry and Warner were educated at the finest
universities, and Mark and Spencer were headed in the same direction. Home was graced with friends, and a certain contentment
that only singleness of purpose between a husband and wife can
create. The running of the ranches, which entailed hard, physical
labor, added to Lemuel's happiness. In a sense, they served as both
training ground and playground for the family.

Whenever he could break away from his busy schedule, Lemuel
loaded his family into the Oldsmobile Touring Car and headed for
the Bar Boot. In those days roads were rutted tracks, and driving to
and from the ranch was an experience. Along the way Lemuel
pointed out grasses, plants and shrubs, and informed his family
what each was good for. Road or not, Lemuel often took off across

the range to view a well, a bull, or a strata of rock. For him, a car had to perform like a cow pony. Luckily, high running boards allowed adventuresome driving.

The entire family enjoyed the ranches. The Bar Boot raised enough pigs for processing hams and bacon in the smoke house, and the O.K. had an apple orchard that the children would ride a horse into and pick apples to take to the cider press near the corral. The fresh apple juice was a delicious treat.

For the Shattuck children the ranches served as an initiation into the sights and sounds of the back country. Sounds were omnipresent: cries of coyotes at night, and creaking windmills pumping water from the well into water holding tanks, which ranch hands referred to as "water holes." Garter snakes often glided on top of the water in these tanks, but that did not prevent the Shattuck boys from diving in for a swim.

One afternoon at the Bar Boot, a cowboy demonstrated to the children how he made jerky. Cutting beef into strips, he placed it into steaming salty brine for a short time, then hung it on clothes lines or fences to dry.

The wood fences strung about the ranches were havens for lizards, which scooted up and down the posts and sunned in the summer warmth. Lemuel often amused his children by obtaining a hair from a horse's mane or tail and making a small lasso. This he dangled in front of a lizard running against the bright sun. Unable to see the lasso, the reptile was always caught. He let each lizard go and allowed his children to try their luck. If they succeeded in snaring one, Lemuel requested it be released, emphasizing he did not believe in hurting or killing any of God's creatures large or small. He related how in his youth he had to kill game for food, but would never kill for sport as so many of his friends did.

Often at night, after dinner at the ranch, the family sat on the veranda and listen to John Meadows tell tales of bygone times. He spoke of the economic panic in Texas following the Civil War, and how he attempted to escape poverty by joining a youthful gang of train robbers. Meadows soon ruled out a career of lawlessness and traveled into New Mexico, where he secured a job as a cowpoke on a small ranch. While at this ranch, he was visited by a former gang member who attempted to enlist him in a projected robbery. That night Meadows quit his job and rode into Arizona Territory to work for the Erie Cattle Company.

Meadows frequently produced an old empty cigar box filled with photographs. One snapshot of four men standing up against a wall with their eyes shut especially fascinated the children. When Isabel asked Meadows why the men had their eyes shut, he replied that the four had committed the Bisbee Massacre in 1883 and had been legally hanged at Tombstone. After they were hung, they were stood up against the wall and photographed.

Juan, a Hispanic who lived on the ranch, usually joined the family on the veranda. As a child, Juan had lived in Bisbee and had broken his leg. The compound fracture was never set properly, resulting in a particularly lopsided limp, a handicap that prevented him from finding employment. When Lemuel acquired the ranches, he compassionately asked Juan if he would like to work for him. Juan was delighted and accepted the offer. A small adobe house was built for him near the headquarters, and a burro purchased to aid him in his chores. He mended fences, picked up dry wood for fireplaces in the main house and performed other small tasks about the ranch. Like Meadows, Juan had a repertoire of stories and superstitions, some handed down from several generations.

In soft Spanish, he told of the abandoned haunted house near the entrance of Rucker Canyon, where a woman's voice could be heard singing *Ave Maria* at midnight. He warned of Skeleton Canyon, where men had lost their lives to Indians and evil white men. It was a bad place to be caught in after dark. If that occurred, the horse one rode would shuffle along as if hobbled, making it difficult to escape the canyon until the sun rose again. Juan's tale of lost gold bullion buried near College Peak east of Douglas was the Shattuck family's favorite.

Everyone in the family, except Isabella, enjoyed ranch food. Pinto beans were always simmering on the back of the wood-burning stove, next to the pot of strong coffee. Ranch cows provided milk and butter, and two hot breads were served at every meal: biscuits and cornbread, along with homemade jams and jellies. Fresh beef was scarce, jerky being served instead. Chicken, however, was plentiful, being served fried, baked, or stewed. There were eggs galore. Cakes and pies topped off all lunches and dinners.

The only time fresh beef appeared on the dinner table was during roundup in late October or early November. A suitable animal was selected and driven into headquarters where it was slaughtered and the meat hung for a few days on a cool screened porch where coyotes or lobo wolves could not get to it. The viscera that would spoil—the heart, kidneys, liver, marrow, gut or tripe—was made into "son of a bitch stew." Though it was tasty, better than steak or kidney pie, most cowboys did not like it. The men instead preferred the simple fare of bacon, beans, jerky and biscuits. If cowboys and guests showed up in large numbers, a quarter of beef would be placed over hot coals in a deep pit, covered and cooked overnight. Served with barbecue sauce, the beef was the favorite roundup dish.

Occasionally the Shattuck family would drive home from the ranch after dark. The trips back to Bisbee were adventures fraught with terror. Summer rains turned the roads into quagmires, which often mired the car up to its axles. Lemuel and the boys dug the car out and continued on to the next bog. Coyotes appeared out of the darkness to run beside the automobile. Ever protective of her children, Isabella feared one of the animals would leap into the car. Lemuel smiled and assured her that these animals, having seen the car lights, were just curious. He was right; the coyotes would soon vanish into the night. A rattlesnake in the road invariably evoked Mother's fear of reptiles. Although Lemuel loved her deeply, he could not help but wish she would accept and appreciate Arizona and its creatures as much as he did. He knew that was an impossible order.

Although five years of age at the time, Isabel vividly recalls one return trip in 1914. On a hot Sunday afternoon as Lemuel steered the Oldsmobile Touring Car through Douglas, they pulled close to the international fence that separated the smelter city from the

Mexican town of Agua Prieta, then racked by battle between government troops and revolutionary forces. Lemuel stopped the car and walked to the international line to inquire as to how the battle was proceeding. A wind from the south wafted an acrid odor about the car, and Isabel complained to her mother. She sat silent until Lemuel returned. Once underway, she asked him about it. The stench, Lemuel matter of factly replied, was of burning corpses. A truce had been called in the fighting and the dead were being burned for sanitary reasons.

Outings were a favorite form of recreation for the Shattuck family. Often on Sundays Lemuel and Isabella loaded their brood into the automobile and drove to the Huachuca Mountains for a picnic in Ramsey or Miller canyons. The latter, being more isolated and primitive, was the family's favorite. While Mother picked watercress along the banks of the mountain stream, the children foraged for wild grapes or berries, which Isabella turned into delicious jelly at home. The fresh, cool mountain water was invigorating. Motoring home, the Shattucks enjoyed the expansive panoramic view of the valley far below, with its numerous creek beds lined with black walnut trees. If the season was right the family stopped in the groves and gathered fallen nuts.

Some summers, before the boys went away to college, the Shattucks boarded a train and went to California to enjoy the seashore and visit Grenfell relatives. A period of intense packing preceded such trips. The heavy trunks were picked up and transported to the depot by draymen. Upon the family's return, horse drawn wagons returned the trunks to the Shattuck home. During the years immediately preceding World War I the family drove to California at least once a year. Two cars would be required for the trip, which was made in stages. The first night was spent in Tucson after a picnic lunch on the way. The next night the Shattucks stayed in Phoenix. One year, a heavy summer storm stranded the family at a Ludlow inn. The haven was crowded and the proprietress gave up her own bedroom to Isabella and the girls. She doled out blankets to Lemuel and the boys, who slept on the floor of the wide central hall of the territorial-style building. At Needles, the family waited for the one-car ferry to haul them across the Colorado River. It took days to reach San Diego and the seashore.

Vacations at the Bar Boot Ranch, outings to the Huachuca Mountains, rigorous trips to California, and frolics along the seashore would soon be cut short by events of great magnitude — events that would splinter the Shattuck family, and divide Bisbee into armed camps.

The directors of the Shattuck-Arizona Mining Company about 1917. First row, left to right: Lemuel Shattuck, Archibald Chisholm, Thomas Bardon. Top row, left: Martin Pattison, Byron Pattison in center, Thomas Bardon, Jr., right. (Fathauer collection)

EIGHT

WORLD WAR ONE

IN January 1914 mining companies of the Warren District were running around the clock, pouring out 7,000 tons of red metal a day—a reflection of the fact that the price of copper stood at seventeen cents a pound. The largest labor force in the District's history—5,000 men—toiled in black, steamy depths; the resultant monthly payroll of $800,000 insured prosperity for merchants. It was a time when gold coins jingled in everyone's pocket. Over three million dollars in 2,000 accounts gave Bisbee bankers something to gloat about.[1] Roominghouse and hotel accommodations were difficult to get, new homes unobtainable. The town was congested night and day.[2] Enthusiasm for the future was evident everywhere.

Six months later an event in far-off Europe clouded this atmosphere with uncertainty. On June 28, 1914, Archduke Franz Ferdinand, heir to the throne of Austria-Hungary, was assassinated by Gavrilo Princip, a nineteen-year-old Bosnian Serb. The Empire accused Serbia of conspiring in the assassination plot, and demanded that the little country root out subversive and anti-Austrian factions in its government and military and suppress publications inciting hatred and contempt for the monarchy. Serbia was given forty-eight hours to comply.

Serbia neither planned nor abetted the assassination of the archduke, and was certainly in no position to provoke a war with Austria-Hungary. In fact, Serbia was willing to accept all Austro-Hungarian demands that did not violate its independence as a sovereign nation. The Empire, however, did not want concessions — its aim was to destroy Serbia and the assassination furnished the pretext. On August 12 the first contingents of three Austrian armies, which would eventually total a quarter-million men, entered Serbia.

The Balkans were the key to southern Europe and thus a pawn in power politics. Any conflict in that part of Europe instantly polarized loyalties. Germany aligned herself with Austria, and their threat to Slav sovereignty immediately drew in Russia. Within the first four days of August 1914, Germany declared war, successively, on Russia, France and Belgium. Great Britain responded on August 4 by declaring war on Germany. On August 5 Montenegro declared war on Austria; and the next day Austria declared war on Russia, and Serbia on Germany. Finally, on August 12 France and Great Britain declared war on Austria-Hungary.

European hostilities generated an outpouring of nationalistic sentiment among foreign-born miners not only in Bisbee, but every American mining camp. Native Austrians and Germans, and ethnic groups sympathetic to Austria and Germany, such as Finns, Croats, and Slovenes, rallied to help their countries. The Cornish, Welsh, Serbs, Montenegrins, Greeks, and Italians responded in like vein, their loyalties slanted toward the Allies.

As early as summer 1914 Bisbee Serbs had raised several thousand dollars for the Serb-American Red Cross Committee. At the same time a company of volunteers was recruited for the Serbian army. On September 28 fifty Serbs departed Bisbee, paying their own way to New York, where they boarded British ships waiting to transport them to the war zone.[3] On July 7, 1915, nearly 200 Serbs and Montenegrins left on a chartered train bound for Chicago. From there they traveled to Canada where they embarked for the Balkans.[4]

Despite every effort by President Woodrow Wilson to keep the United States neutral, the wealth of America, as well as its hereditary ties to England, drew it inextricably toward war. The torpedoing of the *Lusitania* off the coast of Ireland on May 7, 1915, with a loss of 1,000 lives including 128 Americans, instantly crystallized public opinion against Germany. Still Wilson maintained a neutral stance; but when Germany issued a proclamation on January 31, 1917, of unlimited submarine warfare against all maritime commerce — neutral as well as belligerent — the President had no choice but to break diplomatic relations and put the decision for war in the hands of Congress.

On April 2, 1917, Wilson asked Congress for 500,000 men and authority to mobilize the nation's resources. Two days later plans were announced to raise 2,000,000 men through recruitment and conscription. The declaration for war was adopted on April 6, bringing patriotism to a boiling point. "Loyalty Day in Arizona" was declared on April 8 and the Warren District turned out en masse to parade down Main Street in the largest demonstration yet held in the District. Shortly thereafter the Warren District Defense League was formed to undertake a military census and register every male citizen between ages 21 and 30. On April 28 Congress passed, by an overwhelming majority, the Army Draft Bill, designed to raise an army of eighteen divisions.

Blowing mine whistles and firing guns awoke Bisbee on June 5 to a glorious flag raising to celebrate "one of the greatest days in the history of the nation — the day on which every man between the ages of 21 and 30 will register for service." That day 3,000 men were enrolled for possible induction at various precincts of the Warren District. Warner Shattuck, a student of the Minnesota College of Mines working that summer in the manganese ore department of the Shattuck-Arizona Copper Company, registered, as did his brother. Henry Shattuck, a student at the University of Arizona, left his father's farm at Somerton and went into Yuma to register. On July 21 a list of the first 1,400 Cochise County men eligible for the draft was published in the *Daily Review*. The names of both Warner and Henry Shattuck appeared on the second quotas of their respective counties of residence.[5]

Throughout August draft notices went out. The response in homes of draftees is not recorded in newspaper accounts: it was assumed that everyone was motivated by patriotism. Although neither boy had been called, the Shattuck household feared they would be taken before they could finish their education. Regardless of innermost feelings, on the night of September 2 a gala celebration was held in Bisbee's YMCA to honor thirty-four men of the first quota who would soon became "Sammies." Two days later they reported to Sheriff Harry C. Wheeler at Tombstone for induction into the Army.

Concurrent with draft registration came the call for volunteers. Army recruiting officers overran Bisbee. Store windows and power poles were plastered with placards bearing the motto "Uncle Sam Needs You." A week prior to declaration of war twenty-one youths responded to the "Call to Colors" by joining the Navy.[6] Diminutive yet tough Sheriff Wheeler, of Cochise County, applied for officer training, only to be turned down because of wounds sustained as an Arizona Ranger. He would attempt to enlist four more times before finally being accepted. Frank Brophy, a student at Yale, also applied and was accepted into officers' school at the San Francisco Presidio.

Programs to conserve food and natural resources were established overnight. "Wheatless Monday and Wednesday," "Meatless Tuesday," "Porkless Saturday," were grudgingly accepted. Sugar bowls were removed from restaurant tables. When the Red Cross launched its first War Fund drive in the Warren District in mid-June, 1917 Bisbee contributed $22,000. The goal was $15,000.

Within days of America's entrance into the war, legislation was passed on both national and state levels requiring all aliens to surrender their firearms within twenty-four hours to the nearest peace officer. A few weapons were collected by Bisbee authorities. "Enemy" aliens — people from countries at war with the United States — were registered and restricted to zones where their activities could be monitored.

Anti-German feelings manifested themselves as early as June 1914, and reached fever pitch in 1917. Serbs and Croatians throughout America were at each other's throats. When Croatian national organizations voted to give money to the Austrian Red

Cross, Serbians responded with verbal abuse. Bisbee Serbs accused their Croatian neighbors of embarrassing the cause of all Southern Slavs, which resulted in fights. Finns and Irishmen were viewed as subversives. Sauerkraut was banned in restaurants. Native Germans everywhere feared for their lives. Feeling ran so high that dire consequences befell those imprudent enough to speak against the war.

Bisbee streetcar conductor G. W. Strode received two years at Fort Leavenworth for saying that "the president has sold the country to the Cousin Jacks." August Sandberg, a Swedish metallurgist employed by Phelps Dodge at its Douglas smelter, also spent two years at Leavenworth for uttering "aspersions against the President," and stating that "Germany was justified in invading Belgium and sinking the *Lusitania*."[7]

Within two weeks of declaring war in August 1914, Britain placed copper on the conditional contraband list, subject to immediate seizure unless it could be clearly shown shipments were destined for neutral countries. Price of copper, which stood at thirteen and a half cents a pound at outbreak of war, receded to eleven cents in mid-November 1914, paralyzing the domestic metal market. Copper companies everywhere cut production. A thousand men were laid off in the Michigan copper district; in the West, Anaconda, Miami, and Inspiration reduced their labor forces fifty percent or more. At Bisbee, the Copper Queen laid off nearly 500 men, the Calumet & Arizona quartered its force, and the Shattuck-Arizona laid off 150 and ceased production altogether. Wages of miners were chopped ten percent.

Copper is vital to warfare, and the European powers, except Russia, consumed more metal than they produced. Germany, for example, produced less than sixty million pounds per annum and consumed 500 million pounds. France and England each used over 300 million pounds per year and produced next to nothing. The Rio Tinto mines of Spain were the principal European source of copper and how long that source would remain open was anyone's guess in 1914. Europe generally, and the Allies in particular, were dependent on nearly 900 million pounds of copper exported each year by the United States. It was only a matter of time until stockpiles were depleted and Europe again looked to America for strategic metal. That came in mid-1915 when the Allies placed orders in the United States for twenty-five million shell casings, requiring 101 million pounds of copper.

The crisis in American mining vanished with the rebound of copper. By summer 1916 copper stood at twenty-six and a half cents and mines in the Warren District were again running at capacity. That year the Mule Mountains produced over 190 million pounds of copper — a record that would stand for twenty years. The price of copper, however, went still higher, reaching thirty-seven cents in March 1917. Production and exploration ran at a frantic pace. Bisbee enjoyed the pinnacle of its prosperity.

Beneath the camp's bustle, however, lurked factors that would combine with fanatical patriotism and wartime hysteria to bring on a violent confrontation between labor and mine management. Although miners were drawing record-high wages, wartime inflation had deeply eroded the purchasing power of their paychecks. Actually they were no better off financially than before the 1907-09 crash. Complaints, if any, had been quickly stifled in the non-union camp, and so pervasive was company power that no one dared speak for the common miner for fear of losing his livelihood.

In times of industrial peace Bisbee was tranquil; in times of industrial turmoil, however, the atmosphere could quickly become charged with fear of insurrection and lost profits. Under such circumstances men in Bisbee, as in every other mining camp, could be quick to grab the gun. As history has repeatedly demonstrated, the initiative for violence invariably passes to those best prepared.

The Western Federation of Miners (WFM), an organization of "tough, uncompromising frontiersmen," was founded at Butte, Montana, in 1893. Three years later the union affiliated with the American Federation of Labor (AFL), but withdrew in December 1897, ostensibly because sickness and death benefits seemed to be the sole aim of the larger brotherhood. Big Bill Haywood, Vincent St. John, and Charles H. Moyer, the more militant leaders of the

WFM, viewed the AFL as no more than a "coffin society," and hence no spokesman for Western labor. It was to offer a forum to all workers west of the Mississippi that the WFM organized the Western Labor Union in May 1898, and actively began recruiting members regardless of "nationality, creed, or color."

Over the next seven years the WFM, with its affiliate, collided with the crude power of western employers over the issue of the eight-hour day. At Coeur d'Alene, Salt Lake City, Telluride, Cripple Creek, and Idaho Springs, a pattern emerged of strike, then court injunction enforced by armed guards and militia, beatings, and deportations. In Arizona, where miners generally received above-average wages and where mineral extraction was tightly controlled by autocratic companies, unionism was poorly received. Repression and reprisal promptly curtailed most attempts to organize labor.

In 1903 the WFM attempted to establish a local in Bisbee, but that move was not generally favored by miners. Undaunted, the union returned three years later, and in February 1906, John B. Clark and Marion W. Moore, the latter from Globe Local No. 60, successfully implanted Local No. 106 of the WFM in the Warren District. Battle lines were clearly drawn over the question: would Bisbee be an "open" or "closed" camp? Immediately the Bisbee Merchants Association was formed at the behest of mining companies to oppose organizing efforts; on March 15 the miners gathered to vote whether to accept the union. In a referendum vote conducted by Australian ballot, miners turned down union representation five to one, and the copper companies summarily discharged 400 union sympathizers.

A year later the WFM revived its activities in the Warren District and a strike seemed imminent. The mining companies, however, moved fast: in mid-February 1907, they fired 800 men suspected of union affiliation, and established the Bisbee Industrial Association to counteract the WFM. Aim of the organization was to "protect the best interests of the citizens of the community," to "uphold law and order . . . and to reestablish conditions prior to the advent of labor agitation." The WFM was branded "un-

American"; the mines continued to discharge men, particularly foreign-born employees, who were most susceptible to union organizing; and merchants refused credit to miners not loyal to the copper companies. By April, 1,600 men were out of work, laid off because of union affiliation or, as the companies put it, because the mines needed extensive repairs. That summer the WFM called a strike and 3,000 men walked out. Again the strikers were denounced as "un-American" and accused of falling under the spell of "alien agitators," and strike breakers brought in. The WFM painted Copper Queen management as "soulless." By year's end the strike was spent.

Failure of the WFM emphasized the need for more militant support, and that support was found in January 1905 when twenty-seven revolutionaries came together at Chicago, the hotbed of American radicalism, to draft an "Industrial Union Manifesto" calling for "one great industrial union embracing all industries, providing for craft autonomy locally, industrial autonomy internationally and working class unity generally." This "January Manifesto," as it is sometimes called, announced to the world the birth of the Industrial Workers of the World (IWW), or "Wobblies," as Harrison Gray Otis, editor of the Los Angeles *Times*, labeled them.

The IWW was not just an American phenomenon. Their doctrine was rooted in an international movement spearheaded by such organizations as the Transport Workers of England and the French Confederation of Labor, both of which espoused anarchist and syndicalist ideals. The Wobblies, like their European cohorts, "spoke for those who had no voice," for "those displaced or dispossessed by the march of capitalism." According to Patrick Renshaw,[8] the IWW took up the banner of the "submerged fifth; the immigrant and migratory workers, the unskilled, unorganized and unwanted, the poorest and weakest sections of labor." The IWW would focus its greatest effort on the eight million immigrants from south and east Europe. How successful the union was in attaining its goals is conjecture.

No more than five percent of all trade unionists were attracted to

IWW ranks; membership rarely exceeded 100,000 at any time. Between 1905 and 1915, however, at least a million workers were exposed to Wobbly doctrine, which proclaimed an impending class struggle in America. If there was a class struggle, the role played by the IWW was minuscule. Except for a few isolated cases, the union did little to improve the lot of American wage earners or to further world revolution. Because its propaganda advocated slowdown and sabotage, the organization soon came to be feared as a sinister plot, "hatched by foreigners, anarchists and Bolsheviks, against the American way of life."

United for the moment by common goals, late in 1906 the WFM and IWW staged a strike at Goldfield, Nevada, their demands being institution of the eight-hour day and wages of from three to five dollars per day. Although militia was called in, it was the financial panic of October 1907 that undermined the strike and routed the IWW from Goldfield. Miners and Wobblies appeared united on the surface, but dissension over direct action was gnawing at the ranks of the WFM, and by 1908 most western miners had become disillusioned with the Wobbly cause. Syndicalists Bill Haywood and Vincent St. John stood firm, however, and the IWW emerged as a revolutionary, though somewhat shrunken, union.

The IWW was close to withering when yet another labor dispute pumped life into the movement. At McKees Rock, Pennsylvania, the entire 6,000-man labor force of the Pressed Steel Car Company, a subsidiary of United States Steel, walked out over a wage system premised upon piecework. Here was a fray worthy of IWW intervention. The Wobblies waded in and a series of brutal confrontations followed; but tactics employed by state militia, which left eleven persons dead, swung public opinion to the side of the strikers. The steel company capitulated and the IWW won a major victory.

A new spirit of militancy gripped the Wobblies and other strikes followed: in the lumber camps of Montana and California; at the tinplate works in New Castle and Shenango, Pennsylvania; and among farm laborers at Waterville, Washington. But it was at Lawrence, Massachusetts, in 1912 that the union attained its greatest success by securing a five to seven percent wage increase for over 125,000 cotton workers in six states. Unfortunately other events would undo what good the IWW attained.

Led by an individualistic, anarchist western wing that advocated sabotage and work slowdown, the Wobblies took to street corner pulpits to spread word of the miserable conditions under which migratory labor lived and worked. This "free speech fight" focused on the "bindle stiff," that itinerant worker who packed his bedroll between lumber and mining camp. About thirty free speech campaigns were fought in the Far West before a bloody climax was reached on the night of October 30, 1916, in Everett, Washington, where several persons were killed and hundreds of Wobblies beaten and run out of town. After the "Everett Massacre" the name Wobbly became synonymous with "hobo," "tramp," and "bum."

The sudden and unexpected outbreak of World War I in summer 1914 drove a wedge between socialists in America. Strong German and Russian elements of the party favored the Central Powers, because they believed correctly that defeat of the czar would result in revolution in Russia. On the other hand, American socialists viewed the German Reich with suspicion and favored the Allies. The militant faction of the IWW, which included Irish and Finnish miners who hated Britain and Russia for suppression of their national liberation, advocated all-out opposition to the war by every means. Radical doctrine and opposition to war, as well as a substantial membership of "aliens," made the IWW a visible target for super-patriots and weak-minded individuals sucked into the wave of xenophobia and hysteria that swept America after its entrance into the Great Conflict. Every employer who wanted to stamp out unionism fanned the embers of suspicion. Newspapers and magazines painted the Wobblies as a subversive organization — they were the "industrial Ku Klux Klan," the "Bolsheviks," the "pro-German traitors."

The WFM had not given up on the Warren District following their 1907-08 defeat. So long as Bisbee remained a "closed camp," little progress could be made organizing labor in other Arizona mining districts. Union leaders held the firm conviction that as Bisbee went, so would Arizona. As strong advocates of *laissez-faire* business principle, the mining companies were equally inflexible: labor unions had no place at Bisbee.

Change of WFM name in 1916 to the International Union of Mine, Mill, and Smelter Workers (IUMMSW) in no way lessened the militancy and individualism of its members, for the main issue dividing common miner and management had been clearly defined years before: recognition of labor's right to bargain. Early in 1916 the union again set its sights on the 5,000 miners laboring at Bisbee, and by Labor Day of the following year organizers had enrolled an estimated 1,800 members in the Warren District.

Meanwhile the IWW appeared on the Arizona scene, establishing a chapter in Phoenix called the Metal Mine Workers Industrial Union Number 800 and, under direction of Frank Little and Pedro Coria, taking up the cause of oppressed labor. In fall 1916 and early 1917 the Wobblies participated in the ongoing battle for higher wages in Arizona mining camps, recruiting members from among foreign-born miners and Mexican smeltermen. IWW tactics became painfully clear in *Bulletin Number Two*, a publication of the Metal Workers Industrial Union: "We have sunk our shaft and are beginning to drift and cross-cut. We use the best of materials, 'solidarity,' and an understanding of our class interests; when we shoot down, direct action and sabotage seems to be the best powder." Ignoring the war effort entirely and proclaiming only an interest in the working class, the Wobblies agitated and succeeded in staging strikes at Swansea, Chloride, Mayer, Galconda, Ajo, and Jerome.

Government and corporate reaction was predictable. Early in April 1917 Major General John J. Pershing, commanding the Southern Division of the U.S. Army, directed state governors in his jurisdiction to take steps to safeguard "all valuable industrial properties . . . against attack by foreign foe from within." In response, Governor Thomas E. Campbell, as head of the Arizona Council of Defense, suggested mine owners take steps to keep "undesirables" off the properties, and men of alien extraction (particularly from enemy countries) off payrolls.[9] Even before these directives came down from governmental agencies, Thomas Bardon, who had his hands full with IWW agitation at his iron properties in Minnesota, cautioned Lemuel in a series of letters to "watch out for the Mexicans and Germans. They may run over later on and capture mines, smelters, lead mills, etc."[10]

Such caution was late in coming, however. During summer 1916, when southern Arizona was threatened by a very real prospect of invasion by Mexican revolutionary forces under Carranza, militia units had been raised and Bisbee and the border towns fortified. Fences were erected around smelters and shafts, and guards hired to turn away all but regular employees. Even bank securities in Bisbee were moved to depositories in Los Angeles for safekeeping.

As far as Arizona was concerned, spring 1917 was an inauspicious time for a strike. The state was ready. When Wobblies struck camps in Mohave County and at Jerome, they were charged with vagrancy and deported — all with tacit approval from state and federal government, and the blessing, if not aid, from IUMMSW leadership, who did not want to be "dupes" of the IWW.

The IWW filtered into the Warren District to prepare an organizing campaign among the more than thirty nationalities comprising the camp's population. Utilizing Bisbee's city park as a forum, these organizers — outsiders receiving instructions from Salt Lake City and Chicago — distributed literature and attempted to spread dissension in the most receptive segment of miners, Southern Slavs and Mexican surface workers, a sizeable segment of the district's labor force. While they were careful not to incite violence, the Wobblies intimidated men on their way to work, and families of loyal company employees. Nellie Hoy, wife of the publisher of the *Daily Ore*, was warned that the press would be bombed if it did not support the IWW cause. They also imtimidated a number of restaurants into providing free meals. A box of dynamite was found under the Copper Queen Hospital, and small amounts of explosives discovered in other places. Owners of boardinghouses were told to evict all working miners or have their buildings burned or blown up.

It was at the Miners and Merchants Bank, not the Shattuck-Arizona Mine, that Lemuel realized how dangerous the IWW was. While he was seated at his desk, a Slav miner he did not recognize — and Shattuck knew most Serbs in town — came to his office door carrying a small satchel. As he did with everyone, Lemuel invited the man in.

Motioning him to a chair, Shattuck asked, "What can I do for you?"

The miner sat down, placed the satchel close to his chair, and came right to the point. "Mr. Shattuck, you seem to be for the people, but now I think you are a capitalist."

"What makes you say that?" responded the banker. "I started with nothing and worked my way up."

"I see you riding around in a big Marmon with your family," the Serb replied.

At that point Lemuel was distracted by a ticking noise. "What do you have in your satchel?" he asked.

"A bomb," the fellow retorted.

Realizing he had to act quickly, Shattuck said, "Maybe you are right. I have always worked hard, but I am becoming a capitalist. The Bank of Bisbee, down the street, however, is run by men who have been capitalists a lot longer than I."

That satisfied the man. He picked up the satchel, left Shattuck's office, and headed for the Bank of Bisbee. Lemuel immediately called the sheriff and the Wobblie was picked up. Michael Cunningham, cashier of the rival bank, was terrified by the would-be bomber's visit and was a long time in forgiving Shattuck.

During this period of unrest, Lemuel, who had never carried a gun and found hunting distasteful, carried a pistol in his pocket, and his three youngest children were escorted to and from school by a deputy sheriff. Shattuck was not the average mine manager. He had worked underground in his youth, knew the dangers and rigors of hardrock mining. In many respects he sympathized with the plight of the workingman. On the other hand, he had pioneered the development of Bisbee's third-largest company. He had an investment to protect and a responsibility to 1,500 stockholders, many of whom were Bisbeeites, both businessmen and miners.

A man who had a reputation for "fairness and even friendliness toward labor," Shattuck endeavored to sort out the issues threatening to shut down Bisbee's mines. Against Belle's protestations, he attended every IWW meeting in the city park. Standing quietly at the back of the assembly, he listened to numerous speakers, none of whom he recognized. The Wobblies were tanned men, none had the pale countenances of men who worked underground. How could they possibly speak for the average miner, when they were not miners themselves, Lemuel wondered.

For days Shattuck listened to Wobblie diatribe. They derided the flag, many times calling it a "dirty rag." They ranted against government and the war. Boastful of the number of weapons they had, they preached internationalism and depicted an industrial utopia. The Wobblies vowed to carry on the strike until executives were disposed of and the mines controlled by the union. One night a woman warned the assembly, "Our photographer is taking pictures as the men come off shift. All men who are working are marked. When we win the strike and get the jobs, these men will get theirs. . . . Those scabs are marked men." Here were pills Shattuck could not swallow. Without doubt these men had drifted into camp solely to create dissension. He spoke his mind to the *Daily Review*:

I have been following this strike and propaganda closely since it first started, and not missing a single meeting in the City Park or a chance to hear the leaders talk of their alleged grievances. I was forced to the conclusion that these men who started this trouble and are keeping it going with their voices, are purely and simply socialists of the red brand, who are using this area as a schoolroom for their doctrines, and these copper mines and hundreds of miners as their pawns. It is a melancholy conclusion and reflects sadly on the intelligence of their dupes.

I can understand why these striking men can go once or twice to hear these plausible fellows talk, and to laugh at their grotesque companions, wild fierce denunciations, but for the life of me I cannot see how they subsist upon a diet of soap bubbles.

On June 24 the Wobblies elected an executive committee and formulated grievances that were first aired in the city park before presentation to the mining companies. They called for improved safety and working conditions; no discrimination for membership in any labor organization; and in an effort to gain a measure of equality for foreign and Mexican labor, they demanded the sliding pay scale be discontinued and a flat wage of six dollars per shift underground and five dollars above ground be paid. To give an impression of unity between unions, they asserted that these demands originated with the IUMMSW. Several days later demands were enlarged to include abolition of the physical examination required of miners prior to employment, which was an outgrowth of the Safety First Program. The IWW felt the examination was used to weed out union sympathizers. Displacement of men by technological progress was addressed by a demand calling for two men on all pneumatic drills — an issue fought out in the iron and copper mining regions of Lake Superior and northern Michigan. Also wanted were two men to work in raises and discontinuation of blasting during work shifts, both points the Safety First Program had addressed years before. The union demanded abolition of bonus and contract work in an effort to tap wartime profits.

These demands were delivered to mine managers on the afternoon of June 26 by a committee of six men. Grant Dowell of the Copper Queen and John Greenway of the C&A refused to see the wobblies. For the moment at least, Lemuel Shattuck pushed aside personal feelings, invited the men—whom he labeled as "bad ones"—into his office, and listened to their grievances. He made no comment other than to say he would forward their demands to the company president. As soon as they were gone, Shattuck telegraphed the demands to Thomas Bardon. "Don't recognize IWW or yield the control or management of our property or its affairs to any but the US government," the company president responded. "Consult and act in harmony with the mines there."[11] That night all mine managers met at John Greenway's residence, and agreed, in Shattuck's words "to stand pat, as . . . now was the best time to fight it out." They contacted the sheriff's office, and Wheeler agreed to stand by the mining companies.[12]

The managers unanimously felt union demands had no basis whatsoever. They were not an appeal for adequate wages, for the local scale was higher than any other mining camp in the country. They were not against working conditions, for the companies were notable for the care taken of their men, and they were not for union control, for the local union resented Wobbly intrusion, deeming it antagonistic to their interests. Organizers of the strike were no more than floater or "ten-day men, undesirable everywhere." Their demands were "inimical to good government in time of peace, and treasonable in time of war," declared Walter Douglas, president of Phelps Dodge Corporation. "There will be no compromise, because you cannot compromise with a rattlesnake." John Greenway, manager of the Calumet & Arizona, was equally adamant. Lemuel Shattuck, head of the Shattuck-Arizona and Denn-Arizona Copper companies, refused to recognize either the IWW or its demands. If need be, they would close their mines before submitting to what they considered extortion.

All managers, including Shattuck, agreed to continue work, running only during the day. The night shifts would be discontinued "as it woud be dangerous for the miners . . . to come from the mine to their homes in the dark."[13]

On June 27, 1917, the mining companies published their response to IWW demands. Lemuel Shattuck spoke for his company:

The Shattuck Arizona Copper Company hereby notifies its employees and the public at large that demands made by the Metal Mine Workers Industrial Union No. 800 were not accepted or considered.

The management of the Shattuck Arizona Copper Company believes that the Metal Mine Workers Union and its representatives who are the floating population of the district, have not expressed the sentiment of the mine employees of the Warren District who have the welfare of this community at heart.

We know that the conservative mine employees realize that working conditions in Bisbee are better than anywhere in the world and that this is the highest paid wage camp in the United States.

Continual progress is being made for the betterment of working conditions and the management of this company feels that the co-operation of its employees has been the greatest measure of the success of the company.

We believe that the demands of this I.W.W. organization are unreasonable and are the plans of a nationwide conspiracy by enemies of the United States government to restrict or cut off the copper output required to prosecute the war.

The Shattuck Arizona Copper Company hopes its employees will repudiate the action of the union by continuing at their usual occupation.[14]

Most miners did not want to strike, but fear of being labeled a "scab" was persuasive, and on June 27 nearly half of the work force struck. Virtually the entire surface crews at both the Copper Queen and Calumet & Arizona mines walked off the job. Picket lines formed about mine collars, the railroad depot and ore loading docks. The two companies drastically curtailed work. Its isolated location placed the Shattuck-Arizona Company at the mercy of Wobblies. Posters advocating sabotage marked every turn of the road leading to the mine. Lemuel closed the mine, but kept its pumps and compressors going.[15] Strikers did not mince words in letting working miners know what would befall them: "If you go to work tonight you will get picked off," one miner was told. To another, a Wobbly hissed, "If you care anything for your life you will not go out tomorrow night." Another miner was told, "I've got you marked between the eyes." And still another, "You big humpback son-of-a-bitch, we have got you marked." At the same time IWW leaders exhorted their followers to stand firm, for the battle was just starting, would spread across the country and ultimately curtail copper production; fifty thousand harvest hands in the Midwest were ready to leave the wheat fields in a sympathy strike. True or not, Wobbly claims convinced more miners to leave their jobs and unnerved the mining industry generally.

Rumor that the strikers had been infiltrated by pro-German extremists bent on sabotage increased the sense of peril, and Bisbee mustered its strength. The Bisbee's Citizens' Protective League (formed in August 1916 when unions tried to organize the town's clerks and drivers following a labor dispute at the English Kitchen Restaurant) was reactivated to protect business property against vandalism, and placed under direction of Sheriff Wheeler, a soft-spoken, combative ex-Rough Rider and Arizona Ranger.[16] It is fitting to note that Lemuel Shattuck's name does not appear on published rosters of the Citizens' Protective League, although the Miners and Merchants Bank is listed as a member. At time of the founding of the Citizens' Protective League Lemuel was vacationing with his family in southern California.

Reports of weapons and dynamite being cached by Wobblies in the hills surrounding Bisbee intensified the ominous situation, and Wheeler requested that Governor Thomas E. Campbell dispatch federal troops to Bisbee "to prevent bloodshed and the closing of the great copper industry now so valuable to the United States Government." The governor immediately recommended to the secretary of war that the situation be investigated to ascertain if troops were really needed. That request was acted upon. Lieutenant Colonel James J. Hornbrook arrived in Bisbee on June 29, and inspected the town and mines the following day. Reporting to the governor by telephone, Hornbrook stated that "everything was peaceable, with few gatherings of men and no riots." On July 1 Hornbrook reported "conditions slightly improved." If the situation did deteriorate, the officer assured Sheriff Wheeler a detachment of cavalry was stationed at Fort Huachuca, a short distance from town, and the military would extend all possible assistance.

In the meantime, the mining companies endeavored to coax the men back to work. The *Daily Review* emphasized the wartime need for copper and implied the strikers were unpatriotic. When gentle persuasion failed, the companies attempted to scare the miners and spread uncertainty through their ranks. If the men did not return to their jobs, threatened the newspaper, the mines would close altogether. Language was harsher in newspapers elsewhere. On June 29 the New York *Times* blamed the strike on German agents; terms such as "alien," "wobblies," "slackers," and "traitors" appeared in headlines coast to coast.

Although the strike had been relatively free of violence — in fact, labor organizers had cautioned their followers against carrying weapons — the *Daily Review* published a message from Sheriff

Wheeler on July 1 calculated to elevate the level of fear still further:

TO ALL DEPUTIES — I want to impress upon each deputy sheriff the absolute necessity for extreme self-control, cool, calm judgment and patience. Avoid all display of weapons. Remember, you are deputized for protection of self and property and the maintenance of peace. You are subject to my call, a call which will be made when necessary.

Next day more than a thousand miners, loyal to the mining companies, formed the Workman's Loyalty League. Sensing confrontation, 300 miners departed the district, and some strike leaders left for copper camps elsewhere in the state. On July 3 noticeably fewer men were seen on picket lines and more men at mine collars desiring work. Fearing the Wobblies would clash with the patriotic parade of the Loyalty League scheduled for the Fourth of July, Sheriff Wheeler deputized additional men, including Lemuel Shattuck.[17] Except for hisses and jeers, the Wobblies were well behaved.

The strikers were a stubborn lot, united by fraternal loyalty typical of western mining men. They vowed not to return to work until their demands were met. The copper companies likewise toughened their stance, threatening "to take stringent measures upon any sign of disobedience to the regulations which have been fixed by them in association with the Protective Association and the Workman's Loyalty League." Harking back to days of common law that governed mining districts — law that hanged John Heath — the *Review* stated "these combined forces represent the greatest organized strength that has ever been brought about in an Arizona district for the maintenance of law and order."

At that juncture, Charles Moyer, president of the IUMMSW, wired Governor Campbell that his union considered the strike at Bisbee "unauthorized," and on July 6 Moyer revoked the charter of the Bisbee local on grounds its membership had been exploited by the IWW. The *Review* immediately announced Moyer's actions had sounded the "death knell of IWW activity in the Warren District."

Tension eased over the next few days, indicating the strike was "dying a natural death"[18] It was only a lull before the storm, however. Every day the strike dragged on, copper companies and local business suffered enormous losses of revenue. Commerce at Bisbee had taken a dive during the 1907-08 debacle and was just beginning to recover from the plunge of copper prices in 1914. Businessmen were not about to take still another financial beating. Added to the monetary loss was the very real fear of violence. The mining companies, led by the Copper Queen, mobilized to return the Warren District to a semblance of normalcy. In a series of meetings held at the Copper Queen dispensary the first week of July, the top echelon of mine management contrived a plan of deportation — to be used only "in event of a riot." The exact strategy was cloaked with secrecy, for such action had to appear spontaneous. The Bell Telephone office at the Copper Queen Hotel, managed by George F. Kellogg, would be the communication center for the mustering force. A simple telephone message, "This is the Loyalty League Call," would be sufficient to summon both the Citizens' Protective League and the Workman's Loyalty League.

Spontaneity marked the meeting of the two vigilante groups, called on the night of July 11 by Grant Dowell, manager of the Copper Queen mines. Absent were Walter Douglas and Lemuel Shattuck, but John C. Greenway, C&A manager, as well as a number of high ranking mining and businessmen, were in attendance. Sheriff Harry Wheeler chaired the meeting and posed the opening question: what remedies were available to break the strike? There was only one course to follow, asserted Greenway: "Get a train and run the strikers to Columbus, New Mexico, where Uncle Sam would take care of them." Grant Dowell and Dr. Nelson Bledsoe, the latter chief surgeon for the C&A mines, talked in surgical terms about "cancerous growths." Both men thought the malignancy should be removed by an operation.

One mining company executive was not in favor of this course of action. Shattuck believed the strike would wither from lack of support from the camp's resident miners. Demise of labor opposition could be hastened, not by deportation, but by merely firing agitators who had infiltrated the labor force. Although Thomas Bardon, Shattuck-Arizona president, detailed how Wobblies were combated in the midwest: "They arrest them for vagrancy up this way and throw them in jail. When they won't pay the fine, order them out of town." Lemuel Shattuck demonstrated his displeasure of the deportation scheme by boycotting the meeting altogether.[19] In so doing he won the praise of *Dunbar's Weekly*. Lemuel Shattuck "showed his courage and his patriotism by refusing to be a party to the disreputable deportation scheme of the Douglas family," commented the pro-labor journal.[20]

That operation consisted of arresting "on charges of vagrancy, of treason, of being disturbers of the peace of Cochise County, all those strange men who have congregated here from other parts and sections for the purpose of harassing and intimidating all men who desire to pursue their daily toil." A unanimous vote of confidence followed, and the "surgical procedure" was planned with precision of a military movement.

It was agreed that all deputies and loyalty leaguers would wear white handkerchiefs around their arms — like the men of Paris at the St. Bartholomew's Day Massacre — to distinguish themselves from the strikers. Points of rendezvous were established and a course of action outlined for the Wobbly roundup. Weapons were then issued from a small arsenal cached in a back room of the dispensary. Sheriff Wheeler instructed his deputies to assemble at assigned posts at four o'clock the next morning, and upon his cue the town would be swept clean of riffraff.

At 2:00 a.m. on July 12 George Kellogg, aided by extra switchboard operators, began summoning Loyalty Leaguers, some from as far away as Douglas. By six o'clock the task was complete; a *posse comitatus* 2,000 strong huddled in small groups behind buildings, along canyon trails, and on the main arteries leading to town. There were miners carrying pistols, businessmen with rifles, shotguns, and cartridge belts, and a few old frontiersmen with Colts in their belts. All wore white armbands.

Although the plan was distasteful and ran contrary to his philosophy of live-and-let-live, Lemuel Shattuck swore an oath to

Loyalty Leaguers rounding up suspected Wobblies, strikers, and sympathizers on morning of July 12, 1917.

(Courtesy of Arizona Historical Society)

uphold the "duties of the office of Deputy Sheriff." When his phone rang that morning, he too dressed, ate breakfast, and armed himself. Belle tied a white handkerchief around his arm, and Shattuck left the house to take up his assigned position.

At 6:30 Sheriff Wheeler gave the command and deputies swarmed down the canyons, routing dazed strikers and their sympathizers from homes and boardinghouses, from beds, breakfasts with wives and families, and from off the streets. They were prodded like cattle toward the post office and plaza. Within five minutes 500 men had been rounded up. Many were half dressed and most had not eaten. "They began to mill around," reported the *Review*. "The leaders were haranguing them. The moment was critical. Several deputies raised their rifles threateningly and it surely looked as though blood would run on the streets. But . . . the steady tramp of men resounded on the streets leading up the canyon. They were coming down, the reserves two hundred strong with their rifles on their shoulders. The front of the line came up to the plaza and someone called halt. The crash of musket butts on pavement stilled the mutterings and murmurs of the trapped 'wobblies.'... The heart went out of the boldest of the leaders right then...."

In only one instance did a striker offer resistance and that was "quickly and effectively overcome." James Brew, a former miner and IWW card holder, answered vigilante demands to vacate his roominghouse with several shots, killing Orson P. McRae, Mormon shift boss at a Copper Queen mine and a Loyalty Leaguer. McRae's companions promptly shot and killed Brew. The dragnet was thorough. Not only were strikers and Wobblies snared, but also a number of Bisbee professionals and merchants who had extended credit and encouragement to strikers. To Loyalty Leaguers such men were un-American, traitors, "little better than Wobblies."

The group assembled at the plaza grew steadily, until at 7:30 a.m. came the sharp call of "fall in there, march." Armed guards herded the undesirables three abreast toward the Warren ball park, two miles distant. The multitude shuffled along, their numbers still increasing. "What a study in faces as the procession ambled by," commented the *Daily Review*. "Old offenders with sullen brows and smouldering eyes. Foreigners with heavy, stolid looks and bearded, unwashed faces. Young men who looked half-frightened and half-ashamed. Sorrowful, simple, soulless faces passed like a bad dream. . . . Some of the men laughed from pure bravado, but most of them looked very serious. Many glanced furtively, from time to time over their shoulders as though they expected to see the ghosts of old sins. Dejected looking men marched here. Here came a man with a bloody face and his jaw awry; he had offered resistance. Here was a man without a shirt; they had hauled him out of bed. That man with his hand to his head had not raised his hand quickly enough a half hour before and now he walked with his hand raised to caress tenderly a big bump raised by the butt of a colt's army pattern."

The long column snaked its way around the Bisbee depot, passed

(Above) Interning Wobblies at the Warren ball park. (Below) The El Paso and Southwestern train, carrying 1250 strikers, pulling out of Warren and bound for New Mexico.

(Both photos courtesy Arizona Historical Society)

up the tracks toward Warren, entered the ball park, and diffused into the grandstand and onto the baseball diamond. Armed Loyalty Leaguers ringed the field. For hours the men were interned while John Greenway, and others of Bisbee's elite, urged them to recant and return to work. Diehard Wobblies exhorted the men not to weaken, not to break faith with fellow workers.

About 11:00 a.m. a locomotive and twenty-three cattle and boxcars rolled into Warren Station. The internees responded with a "rousing cheer," although the train was nothing to cheer about. Greenway pleaded one last time. He was answered by hoots and jeers — and that sealed the fate of the strikers. Armed men formed a corridor, similar to a cattle chute, from ball park to train, and the prisoners were led single file to the waiting cars. Within an hour cars were loaded, and with 186 armed guards and a machine gun mounted atop the cars, the train pulled out of Warren at noon. At Lee Station, ten miles east of Douglas, the train paused to pick up water for its cargo and to exchange crews. One other stop was made to change engines and then the "Wobbly Special" rolled through the night toward Columbus, New Mexico, while hungry deportees squatted in three inches of manure.

When the train arrived the next day, Loyalty Leaguers aboard were bluntly told by the town constable that Columbus had no room for the "cargo." The train was forced back seventeen miles to Hermanas, New Mexico, where the deportees were abandoned, after being warned never to return to Arizona. Although an El Paso and Southwestern train brought rations the following day, the men were without shelter until July 14, when federal troops escorted them to Columbus, where they were housed in stockades built for Mexican refugees displaced during Pancho Villa's rampage along the border. For nearly two months the deportees were maintained by the federal government; but daily their ranks thinned as men slipped away to return to families in Bisbee or seek jobs elsewhere. By September 8 only 450 men remained, and a month later the camp was disbanded altogether.

Throughout this time, Bisbee remained a fortified town, gov-

erned by a vigilance committee. Deputy sheriffs patrolled all roads
in and out of camp. All trains at Osborn, Naco, and Douglas were
met by armed men to prevent Wobblies from entering the district.
Persons not rated "loyal" who desired to live or do business within
the city had to appear before a kangaroo court. That tribunal inter-
rogated them as to whether they had participated in or sympathized with the strike, and whether they would work at such
places and at such terms as specified by the court. Failure to satis-
factorily answer those questions resulted in immediate sentencing
to the road gang, deportation, or at best, being barred from town.
"We think we have the IWW menace under control, and we intend
to keep it so," wrote Shattuck.[21]

While Bisbee remained isolated, news of what had happened
there was splashed in bold type across newspaper and magazine
pages nationwide. Was the deportation an act of patriotism designed to protect the flow of a vital commodity or was it a violation
of human rights? To answer those questions President Woodrow
Wilson impaneled, within two months of the deportation, a Mediation Commission, headed by Secretary of Labor William B.
Wilson.

The Commission's investigations turned up damning facts for
all concerned—both against labor and management. The strike had
curtailed copper production by a hundred million pounds, but
there was indication the deportation was not solely motivated by a
sense of patriotism: it was perpetuated to break the back of organized labor in Arizona. The findings held that the strike was
rooted in wartime inflation: an expression of labor's desire to share
in profits being derived from abnormally high copper prices. In a
sense the evidence acquitted the IWW of the charge of treasonable
conspiracy proclaimed against them by the copper companies.

"Was the deportation then, the destruction of hundreds of
homes without due process of law, the establishment of extra-legal
kangaroo courts, the hiring and use of hundreds of detectives and
armed guards...inspired by ordinary strike-breaking motives
rather than by extraordinary considerations of patriotism?...Was
the bogey of the IWW raised like the world-enveloping smoke of

the fisherman's copper bottle in the *Arabian Nights* to becloud the real issues?"

Although the Mediation Commission found the deportation to be "without justification, either in fact or in law," it was powerless to effect a legal remedy. It did, however, shift the burden of responsibility for the deportation and attendant acts of violence to copper management. As the vigilante action at Bisbee appeared to have interfered with interstate commerce, violated selective service laws and Arizona statutes relative to kidnapping, the Commission urged President Wilson to take steps toward making such future occurrences a criminal offense under federal law. But in so doing the Mediation Commission became a Pandora's box for the legal profession.

Civil suits were initiated by 272 deportees against the El Paso and Southwestern Railroad and the copper companies. Each individual sued for $20,000, except William B. Cleary, a lawyer who had been deported, who sued for $75,000. A compromise precluded bringing these cases to trial, however, and each married man with a child was awarded $1,250, married men without children $1,000, and single men $500. Simultaneous with the civil suits came a federal indictment against Sheriff Harry Wheeler and twenty-four prominent mining, business, and professional men of Bisbee, including Lemuel C. Shattuck. They were charged with violating constitutional rights of the deportees by conspiring "to oppress, threaten or intimidate any citizen in the free exercise or in the enjoyment of any right or privilege by the Constitution or laws of the United States." The Circuit Court, and ultimately the Supreme Court, ruled these indictments did not fall within federal jurisdiction and should therefore be handled by an Arizona court. Those responsible for the deportation now faced an attack on the state level.

On July 7, 1919, newly-elected district attorney of Cochise County, R. N. French, filed complaints against the Phelps Dodge Company and 224 citizens of Bisbee. According to the *Daily Review*, the warrant read like "a directory of the pioneer residents of the district, practically every man who has taken any active interest in the affairs of the district for the last twenty-five years being included in the charge." It was indeed a "Who's Who," for among those posting bond were heads of the principal mining companies and their assistants, bankers, merchants, and public officials, as well as ordinary townsfolk and miners loyal to their employers. The initial case, however, focused on one man, H. E. Wooton, a hardware merchant and deputy, who, as the complaint charged, was responsible for the deportation of Fred Brown.

Nearly two months were eaten up in preliminary hearings to determine culpability, much to Lemuel's disgruntlement. "Two hundred citizens of Bisbee, of which I am one," he wrote to Thomas Bardon, "have to go to Douglas each day at 12 o'clock to stand trial before Justice of the Peace Jacks on a criminal blanket warrant to see whether or not we will be bound over to the Superior Court. This is, in my opinion, a farce, but we have to just the same, and it looks to me as if we would have to continue going for every day for at least a month or more."[22] After weeks of jury selection, the case ultimately went to trial in Tombstone before Judge Samuel L. Pattee of the Superior Court of Pima County (the Cochise County judge disqualified himself). Outcome of the case, however, was decided the first day of testimony.

The defense, headed by William H. Burges, a distinguished El Paso lawyer, premised its case on the assumption that the economy and lifestyle of Bisbee had been jeopardized by men advocating violence and sabotage; and that under such circumstances "the right of self-defense is perfectly valid in behalf of the community as well as an individual." Such action, contended Burges, was based on the "law of necessity," which originated "in the primal instinct of man to defend his life and what is his own whether it be his family or his property; a law stronger than any other in the world; a law before which everything else gives way when the occasion presents itself, excepting the supreme necessity that comes to a man of sacrificing his own life for family or community or country." Burges further buttressed his case by introducing the maxim "that the safety of the public is the supreme law."

The fear generated within the community by Wobbly intrusion

into an already touchy situation existing between labor and management was reason enough for vigilante action, Burges argued, and had not action been taken violence would surely have occurred against property and employees loyal to the mines. Examples of IWW tactics, philosophy, and publications espousing slowdown and sabotage were offered as evidence that the union was indeed involved in a conspiracy to destroy the capitalistic way of life. Depositions were introduced that had been obtained from the summer 1918 IWW trial in Chicago, which convicted 101 Wobblies, including Bill Hayward and other leaders, of conspiring to violate the espionage law and obstruct the war effort. If that was not damaging enough, a number of miners testified that they had been threatened with bodily harm if they did not join the strike.

Very little evidence was stricken, although the State introduced witnesses who testified that they were not members of the IWW and that conditions were peaceful in the Warren District. R. N. French, the prosecuting attorney, requested that the jury "lay aside any question of self-defense in this case for the simple reason that the evidence does not warrant...considering that subject. There is no self-defense involved."

At conclusion of arguments, Judge Pattee instructed the jury that strikes were permissible and picketing legal even in wartime — in no way did war change basic human rights. "No expression of disloyalty, no treasonable utterances, no failure to measure up to that standard of patriotism that is the duty of every good American citizen could itself justify such deportation." The Law of Necessity, continued the judge, "is a defense when it is shown that the act charged was done to avoid an evil both serious and irreparable, that there was no means of escape and that the action was not disproportionate to the evil."

On the other hand, cautioned Judge Pattee, "if the jury believed that at the time of the so-called deportation there existed a real, threatened and actual danger of immediate destruction of life and property or that the appearance were such as to create a belief to that effect in the mind of a reasonable man and that the defendant and those associated with him honestly entertained that belief . . . then a case is presented which calls for the application of the rule of necessity."

Judge Pattee concluded by summarizing the charge: "Was Brown forcibly carried from this County and State into the State of New Mexico or did he go voluntarily?" It took only sixteen minutes for the jury to reach a decision. "No man could listen to the evidence adduced during the trial," reported jury foreman J. O. Calhoun, "without feeling that the people of Bisbee were in imminent danger, and that if their fears were ungrounded, yet they were apparently real and pressing." The verdict was "not guilty."

Although charges against the other defendants were dismissed by summer 1922, the price had been high both psychologically and monetarily. The cost of defeating the Wobblies was almost $700,000.00, of which $350,000 were attorney fees. Shattuck-Arizona Copper Company's portion was about $45,000.00; the Denn-Arizona barely $400.00. The bill was exorbitant, but Lemuel conceded to Thomas Bardon that the cost could have been far higher and suggested that it be paid:

> If it had been possible to try our cases separately it would have cost us as much as our proportion of this total expense regardless of the fact that we or our organization had nothing to do with the deportation and was probably not liable in any way.
>
> However, we paid our proportion of the expense to keep the I.W.W.'s out of Bisbee after the deportation which I think would probably involve us in the matter.[23]

Bisbee deportation badly tarnished copper company image. Brutal vigilante tactics and the shock of violated human rights pushed from public consciousness the good that paternalism had wrought in the Warren District. The expenditure of millions of dollars to improve living conditions, provide sanitation and recreation, and ensure the safest possible working conditions, were now suspect as being part of a sinister plot to render the common hard-rock miner subservient to corporate interest. When the copper companies voluntarily raised wages and instituted retirement plans following the 1917 strike, they were accused of trying to placate labor and atone for the deportation. Perhaps so, for there followed a period in which defensiveness replaced pride in the constructive changes mining companies had effected in the Warren District.

Of all the mine managers at Bisbee, Lemuel Shattuck bore the reputation of being accessible to his men should they have complaints. His philosophy was that legitimate grievances should be aired and negotiated. He drew the line, however, when outsiders directed from afar entered Bisbee solely to preach anarchy and utopianism, concepts that provided no foundation for a strike. When shutdown was advocated, and a show of unity in the face of threat was deemed necessary, he joined other captains of industry in what seemed a fight to the finish.

Appeal to patriotism, the prime tactic employed to combat the threat, would prove, ironically, a thorn that would slice deep into the fiber of Lemuel and Isabella Shattuck's relationship. Warner and Henry, the two eldest Shattuck boys, were in Arizona during summer 1917 enjoying a break from their university courses: Henry at the Shattuck farm near Somerton in the Colorado River valley, and Warner working in the managenese ore department at the Shattuck-Arizona Copper Company. Initiation of selective service in July threatened to disrupt the education of both youths.

Isabella and Lemuel were not pacifists as some claim. Both felt there was cause for war with Germany over the issue of unrestrained submarine warfare and the sinking of the *Lusitania*. Like many fathers, then and now, Lemuel hoped his sons would be allowed to finish college before entering armed service. Then they could go into the Army as trained engineers. The family filed petitions for exemptions with district draft boards of both Cochise and Yuma counties. As there were only thirteen claims for exemption, none of which applied to the Shattuck boys, their claims were turned down. [24] Appeals followed, and these too were rejected.

Isabella came apart emotionally with failure of the appeals. A premonition plagued her that something would happen to her boys if they went to war. She could not bear to lose any more of her

Henry Shattuck
(Fathauer collection)

children. Lemuel, however, felt that his sons should serve in the Army. Belle's lamentations brought back the loss of their two little boys, when her grief had been inconsolable. Then, too, she was diabetic and her health was failing. Under the circumstances, Lemuel gave in to his wife's pleadings, and together they planned to spirit the youths across the border.

Warner's name appeared on the second quota of Cochise County men to be mobilized in Bisbee on September 18, 1917.[25] When a special train of Pullman sleeping cars departed Bisbee on the night of September 19, carrying 221 men to basic training at Camp Funston, Kansas, forty-nine men were unaccounted for, included Warner A. Shattuck. "They will doubtless be on the slacker list," commented the *Daily Review*. "Failure to report will result in court martial."[26]

Here was juicy news indeed and the paper focused gleefully on the Shattuck name. "Among the forty-nine slackers who failed to report yesterday . . . was Warner A. Shattuck, son of L. C. Shattuck, the well-known Arizona mine operator. Young Shattuck, it was reported, has gone to Mexico. The Sheriff's office turned his name, along with others . . . over to the U.S. Department of Justice yesterday, and all these men will be hunted down by deputy U.S. marshals." Why the *Daily Review* was having a field day with this piece of news is not fully understood, when it had always refrained from publishing news detrimental to prominent mining men. Perhaps it had made an exception because of Shattuck's objection to the deportation.

At this time of national crisis and labor strife, there was no term worse than "slacker." It had been used effectively to combat the Wobbly menace. For it to be applied to the Shattuck name was a source of deep embarrassment for Lemuel and his family. Nevertheless he offered no excuses, and stated his feelings regarding Warner's disappearance. "I would rather that my boy should go to Europe and fight and die, if need be, for his country, than that he should go into perpetual exile in a foreign country branded by his friends as a slacker. I do not know where he is at this time. If I did I think I should go after him and try to bring him back."[27]

Warner Shattuck.
(Fathauer collection)

That was just the beginning. Warner had joined Henry at Somerton, and the two had slipped across the border into Mexico. "It now appears that there are two sons of L. C. Shattuck listed as deserters instead of one," the *Daily Review* taunted.[28]

Warner assumed the name Edward Grover, Henry the name James Clark. Both went to work in the Moctezuma Mining District, near Lampazos, Sonora, for the Bonanza de Plata Mining Company, a gold property their father had a majority interest in. Sonora at the time was torn by revolution and banditry, but the prospect was deemed safe because of its isolation. For more than a year the boys worked in the shallow diggings, writing home frequently. Desiring to see southern Mexico, Henry left Lampazos early in 1919. He traveled southward toward Guatemala, visiting colonies of American draft evaders along the route. For a while Warner received letters from him. Then the correspondence abruptly stopped. Henry did not return.

Fearing for his brother's safety, Warner traced Henry's route southward, searching saloons, brothels, and gambling houses — all places where a young man might encounter trouble. No one had seen him. It was as if Henry had vanished from the face of the earth. Warner relayed the news to his parents, which sent Belle into a frantic state. Lemuel departed immediately for Sonora.[29]

At Lampazos, Lemuel took up the search and again questioned everyone who knew Henry, and many who did not, but were privy to news: miners, freighters, cowmen, and farmers. He turned up only speculation. Some believed Henry had been kidnapped and demand for ransom would be forthcoming. Others postulated he had been killed by bandits for the small amount of money he carried and his body disposed of. Always the tough one in the family, Warner blamed himself and wept bitterly. He should have never left his shy, quiet brother alone. Lemuel tried to comfort his son, for guilt gnawed at him too. Inwardly, he beat himself for allowing his sons to evade the draft.

Business duties, and the impending trial over the deportation, drew Shattuck back to Bisbee. Before leaving Mexico, however, he employed men to continue searching for Henry. Belle was also in a state of anxiety, mentally thrashing herself for sending the boys to Mexico. Her health was not good, and Lemuel worried that grief and self-recrimination would cause her condition to deteriorate further. They were convinced that so long as there was a chance Henry was alive the search must continue.

Throughout early 1920 vague reports of individuals resembling Henry filtered out of Mexico and Central America, mostly brought in by people seeking rewards. A number of sightings fueled Shattuck hopes. As soon as he was free of the deportation trial, Lemuel would travel south and investigate them. Belle vacillated as to whether she should accompany him. Fearful of her health, her husband did not want her to go, but he could not refuse her. Playing it safe, he applied for copies of both their birth certificates to obtain passports.[30] Belle finally made up her mind; and on June 12 Lemuel requested that J. E. James, clerk of the court at Tombstone, change his passport application to read "L. C. Shattuck and wife."[31]

Despite her ill health Isabella boarded the train with her husband on July 3, 1920.[32] Little explanation for this trip was offered. Informing Bardon as to who would look after the Bisbee properties, Lemuel stated that he was "leaving for the east with Mrs. Shattuck who is ill, and I do not know when I will be back in Bisbee, nor, at present, do I know what my address will be.") They traveled down the west coast of Mexico, stopping to see Warner at Quila before pushing on to Culiacan, where they stayed with Russell Grenfell, Isabella's eldest brother, who had recently moved to Mexico after discharge from the United States Army and opened a Ford dealership.

In the relaxed atmosphere of Russell's home — he had purchased the old residence of a "silver king" in the colonial part of town — the Shattucks rested. Belle's health seemed to improve. Russell, who was coordinating the search for Henry, reported nothing new, but suggested contacting several colonies of draft evaders to the south. Following his suggestion, the Shattucks traveled on to Guatemala, El Salvador, and Costa Rica. Staying at small, often unsanitary inns, they sought out men who had evaded the draft.

They questioned a number of Anglo-Americans teaching English and working at other jobs, but none had encountered Henry or heard of him. [32]

After months of searching, during which time Belle's health declined, they boarded a freighter carrying fruit, passed through the Panama Canal, headed up the east coast of Mexico to New Orleans, and took a train back to Bisbee. Late in December, nearly three months after their return, an old mining friend, E. C. Anderson, reported seeing Henry in an automobile in Guatemala. Dubious, Lemuel followed up: "As I have not heard from Henry for a long time I did not know where he was, and his mother is worrying a great deal about him and his whereabouts, and we would very much like to get into communication with him, if possible. Probably you were mistaken in the identity, as you say you saw him riding in a car. Please advise me whether it was in November last, or a year ago November when you saw him." Like all the other reports, this one amounted to nothing. [34]

Quiet, gentle Henry was gone. His disappearance hung like a shroud over the family. No doubt Lemuel and Isabella knew more about his disappearance than was explained to younger Shattuck children. A strict code of silence was maintained. Henry's disappearance was mentioned only in correspondence to individuals who reported having seen him, and then only in the briefest sense. Never was he mentioned to relatives in Pennsylvania and California, or to close friends. This is understandable in light of the fact that Bisbee had already feasted on the Shattuck dilemma. Silence would protect the family from small town cruelty that would surely result if the fate of Henry leaked out. Having one son still an outcast in Mexico, his guilt fast turning him into an alcoholic, dampened family spirits still further. Gone forever was the gaiety of family get-togethers. Isabella lived the rest of her life not knowing whether Henry was alive or dead, and never gave up hope. Her health rapidly went downhill. Always sternly realistic, Lemuel believed Henry dead. Years later, he confided to his children that he had made the greatest mistake of his life when he gave into Belle's desperate attempt to save their two eldest sons.

Brewery Gulch in 1906. The Muheim Building, housing the stock ex-change of Duey and Overlock, is at center.

(Courtesy Arizona Historical Society)

EMERGENCE OF THE
SHATTUCK-DENN COPPER COMPANY

THE BATTLE against Wobbly intrusion into the Warren District left the mining companies short of experienced help, facing a declining labor pool due to army conscription. Of the total of 591 men employed at the Shattuck-Arizona and Denn companies, 153 were deported. The net result was an 85 percent decrease in production in August 1917.[1] It would be months before output equaled pre-strike levels; by then lucrative copper prices had vanished.

To stabilize the price of metal, the federal government fixed the price of copper at 23½ cents a pound early in 1918. The move curtailed profiteering but hurt American mining companies. When increased costs of labor, supplies, maintenance, and innumerable other expenditures were factored into copper production, this price gave the same return to producers as did thirteen cents a pound in 1914.[2] By fixing the price of copper, mining companies moaned, the government had inadvertently restricted exploration and development. That was certainly the case with companies producing low grade ore.

So far as the Shattuck-Arizona was concerned, it held its low grade ore in reserve, sent only refractory carbonate to the smelter, and turned its attention to metals other than copper: manganese, lead, and silver. The company's Leo and Iron Prince claims contained a million tons of cerussite, a carbonate of lead, as well as cerargyrite, a silver chloride also called horn silver. Both minerals contained high values in gold.[3]

With a ten-year supply of lead-silver ore, the Shattuck-Arizona moved to install its own processing facilities. This time there was no vacillation as had happened years before with the proposed smelter. Lemuel commissioned the Salt Lake City Station of the Bureau of Mines to conduct experiments to determine if its lead carbonate ore could be commercially treated via flotation, a process introduced to Arizona in 1915 by the Inspiration Company. In a series of tests, metallurgists O. C. Ralston and Glen L. Allen formulated a procedure whereby lead carbonate was sulfurized, which allowed silver chloride and lead carbonate to be lightly coated with a film of sulphide, rendering it amenable to concentration by oil flotation.[4] A license to use the process was obtained from Mineral Separation Company of Salt Lake City, which held the patent, and pneumatic flotation machines developed by J. M. Callow were obtained to equip a 400-ton mill to be erected on Denn-Arizona property.[5] The mill was operational by summer 1918, supervised by Glen Allen.

That year the Shattuck-Arizona concentrated more on production of lead-silver than copper, with development distributed through nine levels of the mine. By holding most of its copper off the market, the Shattuck-Arizona saw its 1918 income decline to $2,663,082. Net profit was only $784,658. Factors beyond control of Lemuel Shattuck and the company's directors, however, were working to force closure of the mine. The armistice in November ended World War I, and the government removed its support of copper in early 1919, leaving the mining industry to contend with a worldwide glut of metal. When the price of red metal dropped to eighteen cents a pound, mining companies ceased operations, signaling a slump that would last three years. If that was not bad enough, Martin Pattison, who helped Lemuel pioneer development of the mine, died on December 20, 1918, after a short illness.[6] Six weeks later the Shattuck-Arizona was nearly lost to fire.

The oldest portion of the mine, between the 700- and 800-foot levels, was an area of massive copper and iron sulfide deposits.

Stope No. One in that section produced a large percentage of the mine's copper, leaving a great cavity. To stabilize formations about the hole, the stope was back-filled with pyritic material that generated considerable heat upon decomposition. In a non-oxygen environment the material was non-combustible. The Shattuck Mine was veritable swiss cheese, full of holes both man-made and natural, admitting air currents impossible to close off. Indications that something was wrong came with whiffs of smoke and gas on February 14. Fire crews attempted to reach the burning section, but sulphurous smoke drove the men back. On February 22 mining operations were discontinued. The seventh, eighth, and ninth levels were sealed with bulkheads to choke fire. Air seeped through fractures and the stope continued to burn, throwing off immense clouds of poisonous gas.

If cutting off ventilation would not halt the fire, water would, and Lemuel Shattuck ordered the mine flooded between the 700- and 900-foot levels. Eighteen million gallons of water were pumped into the levels, without success. On March 1 Shattuck reported the fire still burning, "making much gas." He believed the last hope of saving the mine lay in flooding the workings up to the seventh level, which would require forty days of pumping. He telegraphed company directors requesting authority to continue pumping. "Use every means to save the mine," they wired back. Shattuck ran the pumps at full capacity; 350,000 gallons a day.

It was slow work. The mine was full of stopes, drifts, and innumerable cavities. Shattuck and his men kept at it, and on March 14 he optimistically reported to Thomas Bardon that water "is now about four feet above the 800 station level, and . . . it looks now as if we are going to be able to flood the mine and put out the fire." By end of May water stood within five feet of the 700-foot level; the fire was out, and pumping ceased.[7] As a precautionary measure water was allowed to stand for another month.

Water seeped away at a rate of eight feet a day; and in September the mine was pumped and cleanup operations commenced. Cost of fighting the fire was staggering: $140,000. Loss of production was far higher. That year the Shattuck-Arizona Copper Company posted a loss of $186,499, the first in its fifteen year history. Stockholders nevertheless were paid $262,500 in dividends, drawn from surplus profit. [8]

Despite the fall of metal prices, the mine was repaired and put back in production, strange in light of the fact that the Shattuck-Arizona had always closed when faced with slumping commodity prices. Not this time, however. The company had been hit by a federal government tax levy of $640,000. Apparently at end of 1916 the Shattuck-Arizona had over five million pounds of copper due from refineries the company had previously sold. Although the metal was delivered in 1917, the company posted the sale as of 1916. The government, however, viewed the sale as occurring in 1917 and assessed additional tax. [9]

Anticipating a costly suit, or enforcement of taxation, the company continued operating. New ore bodies were opened, one of which was a hundred feet long and averaged six percent copper. By fall 1920 more than two million pounds of copper, eight million pounds of lead, 400,000 ounces of silver, and almost 5,000 ounces of gold had been produced. Except for gold and silver, metal prices steadily declined throughout the year. One by one mines closed, and on October 29 Lemuel ordered cessation of mining at the Shattuck. [10] A crew of sixty miners was kept at work in development, but they too were dismissed in August 1921. The pumps were pulled, and the mine idled.

Although the Shattuck-Arizona was an old mine in 1922, it still was far from depleted, its longevity due to Lemuel Shattuck's insistence on shipping only high-grade ore when price of copper was at its maximum. Such conservative management prolonged the mine's life far beyond that of similar operations, generating maximum profits allowing the company to turn its attention to development of its sister mine, the Denn-Arizona. "Old Shattuck is a wonder both as a man and a mine," Bardon affectionately wrote Lemuel on May 29, 1921.

When it came to richness, the Denn Mine stood in sharp contrast to the Shattuck. Unlike its sister mine, the Denn's ore was buried a thousand feet below the Morita conglomerate, and development between 1905 and 1917 was disappointing. Not only was the Denn plagued by flooding, ore shoots in its successive levels seldom broadened into large bodies of ore. What ore was shipped never covered expenditures, which were drawn from the Shattuck-Arizona treasury.

Location of the Denn Mine was favorable for tapping large masses of copper ore, for it abutted the Dividend Fault, the main geological feature of the Bisbee Mining District. Other properties had proved that mineralization occurred where cross faults joined the Dividend, and at least one such fault, possibly more, existed in the Denn property. Lemuel Shattuck was too much a mining man to throw in the towel just because large ore shoots had not yet been encountered. He gambled on two factors affecting the future of the Denn: deepening of neighboring properties would drain the Denn and allow its sinking to greater depth, and a hike in price of copper to a level where profit could be turned on ore of lower tenor.

The gamble began to pay off when the Calumet & Arizona Company acquired the Junction Mine in 1913 and began deepening the shaft.[11] By summer 1914 the great bore had been sunk and concreted to the 1,800-foot level, where a giant station was cut and pumps installed, lessening the flow of water into the Denn, but not enough to allow resumption of work. One hundred fifty feet of water still remained in the shaft. Over the next two years the Junction Shaft was deepened another 200 feet, fully draining the Denn to the 1,800-foot level.

Draining the Denn was an opportunity for Calumet & Arizona management to turn a profit. John Greenway and Gordon Campbell proposed that a drift, to serve both as a drainage channel and an avenue for mine development, be extended into the Denn from the 1,700- or 1,800-foot level of the Junction, and a fee charged for water drained through the conduit. Additional drifts could be run into Denn ground to facilitate exploration, and ore lifted to the surface through the Junction shaft.[12] Development of the Denn would be conducted "in harmony" with the C&A. In harmony meant an increase in Denn Company stock from 350,000 to 500,000 shares — enough to give the C&A crowd a voice in management of the Denn.[13]

Shattuck was not against C&A people acquiring Denn stock — he could not prevent it once the stock was listed on the market. He recognized, nevertheless, the pitfalls of such an agreement. "Of course, it would be a good plan to work in harmony with them," he wrote Martin Pattison, ". . . but the proposition to pay for all water that might flow from a development drift . . . I do not believe would be a good idea for us, for if their prospect drift was run in from the 1,800-foot level nearly all the water would come from this drift and we would have to pay for unwatering the whole district." Instead, Lemuel proposed to pay for pumping on a tonnage basis of ore mined.[14]

A pumping agreement, according to a tonnage basis, was worked out in May 1917, and on June 20 the Denn-Arizona's articles of incorporation were amended to allow an increase of 150,000 shares of stock at $10 par value, 40,000 shares of which were purchased by C&A management. Flushed with $1.5 million, President Thomas Bardon authorized Shattuck "to begin active operations at once to open up the Denn Mine and place it on a dividend-paying basis as an active shipper."[15]

News of the Denn's revival swept the mining world. Its stock, which had slipped to two dollars a share in 1910, rose to nineteen dollars by summer 1917, when the company shipped its first ore. Erection in late 1915 of a new general office building for the Shattuck-Arizona on Denn property in Johnston Addition, as well as construction of the Shattuck concentrator, gave a sense of permanancy to the Denn. Wobbly intervention during 1917, however, slowed production, and government fixing of the price of copper sharply cut profits. The ending of the War late in the following year induced a general slump in commodity prices.

The steady decline of copper forced Shattuck to consider curtailment of mining. "I cannot see any use producing copper at present price for metal, as we can do so only at a loss," he wrote early in 1919 to Bardon. "Wish you would confer with other directors and get their ideas and advise me so I can cut both mines down to a development basis until conditions change."[16] The board concurred, and on March 17 ordered Shattuck to "curtail all operations at Denn to development work only."

Throughout 1919 a skeleton crew of veteran miners labored in the Denn tracing faint streaks of ore. It was difficult and dangerous work. One drilling crew on the 1,700-foot level encountered water under such pressure that, when the drill was withdrawn, the flow spurted twelve feet from the face of the drift. The water was left to flow for twelve hours in hope that it would lessen. It did not and the hole was plugged. Another watercourse was encountered on the same level. It too was plugged, but seepage occurred throughout the level, much to the worry of Shattuck. "If the flow does not gradually decrease I am afraid that we will lose the level as our pumps are handling over 1,500 gallons per minute, and this is all they can do, as we are forcing them now," he wrote H. L. Mundy, vice president of the company.[17]

As difficult as exploration was, new ore shoots were a regular occurrence, and the mine looked more promising with each passing day. One large mass of copper sulfide in particular, was discovered laying astride the Denn and C&A boundary — a situation calling for "working in harmony." Stoping of that body would wait until water could be drained through the Junction Mine. Considerable ore reserves were now in sight, but the water problem ate voraciously at the company's cash, and worked against rising additional revenue.

To finance development of promising ore streaks on the 1,600- and 1,700-foot levels, the company's directors proposed turning over a minimum of 100,000 shares of stock to Hornblower and Weeks for listing on the New York Stock Exchange. Lemuel conceded Denn would develop into a large and valuable mine, but advised caution in selling stock until flooding had been controlled:

If Hornblower and Weeks boosts the price of the stock up, and sold same, and we should be drowned out for any length of time, it is my opinion that the Denn would then take quite a relapse and it would afterwards be hard to get the confidence of the people. I think we should not list the stock nor give Hornblower and Weeks a call on any of the same until such time as we get on a steady paying basis.

On Shattuck's advice the directors did not plunge into a stock sale at this critical time. Instead, funds to develop the Denn were drawn from the Shattuck-Arizona treasury — a risky procedure in light of slumping metal prices. Not wanting to tap unsold copper reserves amounting to over four million pounds, or $477,000 in Liberty Bonds drawing four percent interest, Lemuel allowed the company additional credit with his bank. By February 1920 the Denn owed the Miners and Merchants Bank $315,000 and the Shattuck-Arizona Company $156,000. The tax levy against the Shattuck-Arizona compounded the dilemma, finally forcing the sale of its copper below the price of eighteen cents a pound. A number of notes, including those held by the Miners and Merchants Bank, were refinanced with Eastern institutions.[18] Regardless of worsening economic climate, Lemuel kept the Shattuck-Arizona working mainly to subsidize Denn development, borrowing powder and other supplies until word came down from the directors to cease operations. On July 18 management voted to close down "until conditions make it possible to produce a profit." The lead concentrator was kept running until December, and then it too was shut down.[19]

Neither the Denn nor the Shattuck produced ore in 1922. It was just as well, for Thomas Bardon, long-time friend of Shattuck and Pattison, died at end of January 1923, leaving the leadership of the companies in disarray.[20] His death interrupted overtures to merge the Denn Mine with the Calumet & Arizona on the basis of four shares of Denn stock for one of C&A, a swap that Shattuck and most majority shareholders were in favor of. Gordon Campbell, C&A President, bluntly turned the offer down, however, stating that "developments of Calumet & Arizona interests will rapidly widen the differential in valuation" of the two properties.[21]

As the new head of the Shattuck-Arizona Company, Lemuel brought the Shattuck Mine back into production when price of copper edged upward in February. By practicing utmost economy in reducing overhead, close sorting of ores, and reclaiming of supplies, the company was able to produce three million pounds of red metal at under seven cents a pound, none of which was sold in 1923.[22]

Diversity of Shattuck-Arizona ore enabled the company to turn a profit despite the sluggish economy. In late 1923 Lemuel cut back copper ore production by fifty tons a day, and concentrated on extraction of other metals. The lead stopes, in heavy and caved ground on the 600-foot level, were cleared and reinforced. Rich, lead-silver carbonate was hand sorted and sent directly to the El Paso smelter, where it was mixed with ores from Ahumada, Mexico. Lower grade siliceous lead was sent to the Copper Queen smelter at Douglas. In 1923 the company produced over five million pounds of lead which, added to ores having high gold content, covered all but $10,000 a month of operating cost.[23] Lead-silver was the ace up the Shattuck-Arizona sleeve — so long as lead remained at 6.75 cents a pound, the Shattuck-Arizona could remain operational and conserve its copper resources.

Lemuel pushed lead-silver production throughout 1924, and in spring 1925 reactivated the lead concentrator, putting it in charge J. H. Stanley, who had worked for Phelps Dodge at Tyrone, New Mexico, and for the American Smelting & Refining Company at Parral, Mexico. Under Stanley's guidance the concentration process was speeded up, the sulfurized lead carbonate being sent directly from ball mills and classifiers to the flotation plant, thus eliminating tabling.[24]

Sadly, this lead production represented the last gasps of the Shattuck-Arizona Copper Company. While not totally depleted, the mine contained little ore of commercial value under 1925 technology. On the other hand, the Denn had great potential. It had over three miles of workings, tapping numerous ore shoots below the 1,200-foot level. One enormous ore body extended from the 1,250- to the 1,800-foot level, and probably went deeper. Hampered by flooding, the mine produced only a half million dollars' worth of copper up to 1920. With $1,100 a day being expended on fuel alone to drive the pumps, no profit existed.

With each passing month the Junction Mine drove its shaft deeper, reaching the 2,200-foot level in early 1925. It was only a matter of time before a drift would be driven into Denn territory, eliminating flooding to that depth. The Denn's future was indeed bright. That spring directors of both the Denn-Arizona and the Shattuck-Arizona companies moved to consolidate the two enterprises.

A merger would be effected through formation of a new corporation that would combine 350,000 shares of Shattuck-Arizona stock and 450,000 of Denn, all at equal value. A wave of protest immediately arose from Shattuck-Arizona minority stockholders who knew nothing about mining. The Shattuck Mine is "a fully developed and profitable operation; the Denn partially developed and devoid of operating funds," they screamed. "How could a share of Denn stock be equal in value to a share of Shattuck?"[25] As general manager of both companies, the task of conciliator fell to Lemuel Shattuck. He wrote to each disgruntled stockholder, explaining that they were not being short-changed. The best course "to perpetuate the life of the Shattuck Company" was to form a new company and exchange stock so that ample funds could be raised to develop and work the Denn. Hopefully, a new mine could be developed that would pay as much in dividends as had the Shattuck-Arizona.[26]

Despite a lawsuit to block the merger, the Shattuck-Denn Mining Corporation was organized in May under the laws of Delaware with an authorized capital of 1,000,000 no-par-shares. The corporation took over the companies through an exchange of stock, share-for-share, for the outstanding 350,000 shares of Shattuck-Arizona Copper Company and 450,000 shares of Denn-Arizona Copper Company.[27] Officers of the new corporation were the same as those piloting the original companies. L. C. Shattuck, the driving force and manager of operations of both mines, was president. H. L. Mundy, Byron Pattison, and Thomas Bardon, Jr., were vice

The Denn Mine in 1982. (Westernlore Press photo)

presidents. A. M. Chisholm served as treasurer, Norman E. La Mond, secretary, and T. O. McGrath, general manager. The directors were L. K. Baker, Maurice Denn, L. C. Shattuck, Thomas Bardon, Jr., H. L. Mundy, Byron Pattison, A. M. Chisholm, John G. Williams, and R. W. Higgins.[28]

An injunction against the merger filed by minority stockholders dragged through Minnesota courts for a year. Denied by the Duluth District Court, it was appealed to the state supreme court, which sustained the decision in November 1926. Court battles in no way slowed development of the Shattuck-Denn. An erratic streak of altered limestone "thickly peppered with gold" was discovered during summer 1926. The find was impressive. A powder box full of such specimens easily fetched several hundred dollars, and some miners were caught highgrading.[29] A larger compressor plant was installed, and the Denn shaft sunk to the 2,000-foot level. Drifts along the 1,400-, 1,600-, and 1,800-levels cut both carbonate and sulfide bodies; near the Junction side lines a mass of chalcocite was penetrated. This copper ore paid operating expenses while the shaft was deepened and enlarged to four compartments.

In the process of sinking the shaft, the bore pierced the Dividend Fault, entering heavy, wet ground, necessitating installation of jacket sets in addition to regular shaft timbers to permit continuous hoisting and drainage. Sinking was completed and cages put in operation early in October 1927, and drifting commenced at a rate of eight feet a day to explore territory under the ore body found at the 1,800-foot level. By January 1928 five drifts radiated from the Denn shaft, and several were extended from Junction territory into Denn ground.

The four-compartment shaft was completed August 15, 1928, and a concrete collar poured. Because the old hoist was inadequate for the greater depth, $100,000 was expended for new hoisting equipment.[30] An eighty-five foot high, steel headframe was erected by the Kansas City Structural Steel Company, and a Norberg high-speed electric hoist, capable of moving 750 tons a day, installed a month later. Shaft stations were then cut on the 2,100- and 2,200-foot levels, and an agreement was signed with the C&A to drain the Denn through drifts connected at the 2,000- and 2,200-levels to the Junction shaft.[31] Drainage allowed rapid exploration. Drilling crews drifted at the rate of fourteen feet a day on the 2,100-level; ten feet a day on the 2,200, all with good results. Five percent ore was struck. It broadened into a sizeable body. Another ore mass, assaying nine percent copper, was hit.[32]

This exploration proved ore in commercial quantities existed in the Denn, and Lemuel moved to open as much ground as possible. To do so, he employed the conventional method of core sampling, as well as an unorthodox geophysical technique of plotting geologic structures by taking magnetometer readings of magnetic intensity of rocks. Welton Humphrey, and his son Welton, Jr., had attained good results by this method at Signal Hill, California. When they suggested testing their exploration techniques on Phelps Dodge ground at Bisbee, they were scuffed at. Copper Queen geologists grudgingly conceded that a magnetometer might work with petroleum-bearing structures, but such an instrument could never differentiate sizable ore bodies from highly mineralized ground.

More open minded than Phelps Dodge management, Lemuel Shattuck listened when the Humphreys proposed testing Denn ground. In Southern California they had been able to tell the difference between oil-bearing strata and barren ground to a depth of 8,000 feet. While some difficulty might exist in determining commercial bodies of ore in highly mineralized ground, the Humphreys pleaded for a chance to test their equipment and theories. Lemuel had nothing to lose and everything to gain, and he consented. The three men determined to probe and map forty-five acres of Denn property at the intersection of the Mexican Canyon and Dividend faults.[33]

The Humphreys parked their drilling truck on Denn property, and hauled out their electronic apparatus. Throughout spring of 1929, they drilled a pattern of holes into which they lowered what Shattuck called a "doodlebug." Readings from this sensor were painstakingly plotted. This mapping indeed revealed differences between the axis of just mineralization of no commercial value and the axis of what might be commercial ore. To prove these observations, Shattuck ordered a drift run toward the Humphreys' drill holes from the 2,100 level. In June miners encountered ore assaying 33 percent zinc, and 10 percent lead. The ore also carried three ounces of silver to the ton. The "doodlebug" worked. In fact, the instrument had such merit Shattuck recommended the Humphrey's to other mining companies envisioning geophyical exploration.[34]

The Denn Mine moved into production in January 1929, sending 350 tons of ore daily to the Douglas smelter. Even the old Shattuck Mine—written off by geologists several years previous as being depleted—was rejuvenated by discovery of a four-foot-wide streak of forty percent copper in an unexplored section of the 400-foot level,[35] which furnished 100 tons of ore per day. During summer 1929 the two mines produced three million pounds of copper, which sold for 17.60 cents a pound. This production earned the Shattuck-Denn the distinction of being the only company in the Warren District to markedly increase production that year.

The ore streak on the 400-foot level pinched out in summer of 1929 and life of the Shattuck Mine ebbed. Lemuel discharged forty men, reducing overhead $400 a day, yet continued diamond drilling in previously productive territory for another four months. When nothing of value was encountered, he sadly admitted "the Shattuck is nearly played out." He let the drilling crews go, and turned the property over to leasees.[36]

Mining is a cyclical business, tuned to the world's economic climate, which dictates rise and fall of metal prices. In early 1929 companies everywhere were making money. The future looked bright, and vast sums were allocated for development. The Calumet & Arizona deepened its Campbell Shaft to 3,600 feet, and Phelps Dodge moved to raise running capital by introducing its stock to the investing public. That summer no one could predict the Great Depression. On Black Thursday, October 24, 1929, the good times ceased with crash of the stock market. Bisbeeites helplessly watched copper—the mainstay of their prosperity—retreat from twenty-three to eighteen cents a pound. Mining stocks plummeted; Shattuck-Denn dropped to three dollars a share. Fearing that it might vanish from the market, Thomas Bardon, Jr., purchased all the stock he could at four dollars a share, and advised Lemuel to do the same. Shattuck immediately placed an order for 2,000 shares at five dollars a share. Other Bisbeeites responded in kind. By November orders for 10,000 shares of Shattuck-Denn had been placed through local stockbrokers, stabilizing, at least for the moment, the company's securities.[37]

The market retreated further as companies unloaded metal reserves. On April 15, 1930, metal dropped four cents more, "marking the most erratic break on record" for the commodity. The precipitous decline was labeled "merely an adjustment," and brokers tried to calm fears by telling their clients that no immediate change in selling price was contemplated. They were wrong.

Copper steadily declined toward twelve cents — the level between profit and loss for all producers. When it broke the twelve cent barrier in November 1930, mining companies quartered production. The Shattuck-Denn was caught in a dangerous financial position. Throughout 1929 and 1930 it had opened a sizeable ore body in soft, unstable ground. Raises and stopes had been driven into it at every favorable point of attack. Ore production soared to 600 tons daily. So mechanized were the Denn's extraction techniques that the company could mine the remaining 200,000 tons of ore at two dollars a ton. If the company discontinued operations, loose ground would cut off access to this ore, rendering mining of the ore body "almost prohibitive, considering the tonnage left." The hazard of fire also reared its head. If spontaneous combustion occurred in the gob of a sulphide stope, the C&A Company would be forced to bulkhead all opening to their ground, cutting off ventilation, and as Shattuck expressed to Bardon, "we would never be able to get this ore, and probably never be able to again use the Denn Shaft."[38]

To shut down the Denn would be disastrous to the company. Bardon concurred with Shattuck: "The most important job is to explore the Denn . . . especially in the section south and east of the shaft."[39] How to do that in the face of a declining market was the question that absorbed the attention of both Shattuck and Bardon. Both men were painfully aware the Shattuck-Denn was cash-strapped. Only $325,000 existed in its treasury. Another half million dollars was tied up in Liberty Bonds, which could be borrowed against. The largest asset was 4,500,000 pounds of unsold copper. As long as copper hovered at twelve cents a pound, the Denn and Shattuck mines could produce monthly about 600,000 pounds of copper and $8,000 worth of gold and silver, more than enough to offset the cost of development.

The price of copper moved lower, forcing Shattuck and his associates to seriously consider closing the Shattuck Mine, the least productive of the two mines. There was one glaring hitch — closure would set off a wave of selling by jumpy stockholders, undermining the value of Denn securities. To combat that, Shattuck suggested the directors buy stock and stabilize the price at four dollars a share. It was agreed that 27,000 shares would be purchased at four dollars a share. Orders went out, Shattuck purchasing $32,000 worth, and other directors and their relatives placing large orders with brokers.[40] Stock poured in from all parts of the country, in such volume as to easily fill the quota, if not exceed it. What began as a move to protect company securities now threatened to undermine the stock value. Orders were canceled, and Shattuck and his associates adopted a cautious approach. To test the market, they waited several days and then placed orders to buy 500 shares at $3.75, 1,000 shares at $3.50, and a 1,000 shares at $3.25. To keep the Shattuck-Denn running in the meantime, the Miners and Merchants Bank loaned the company $100,000, taking as collateral copper certificates for 2,000,000 pounds of metal.[41]

The fate of the Shattuck Mine was sealed when copper hit ten cents a pound the first week of October 1930. "We do not want to sell our metal so cheaply," Bardon wrote Shattuck. "We believe that it is part of wisdom to prepare for two or three years of hard times in the copper business, and if we are to survive such a period we must begin reducing our expenditures as much and as quickly as possible." Bardon added the postscript: "I hope some way can be found of taking care of the old men at the Shattuck, like Nicholson, Graham, Hogan, etc. They have been with us a long time."[42]

The Shattuck Mine was closed, but diamond drilling continued at the Denn through 1931, revealing an ore body below the 2,200-foot level in the area east of the Mexican Canyon Fault, close to a hole drilled by the "Doodlebug" men. In March 1931 the company sold 675,000 pounds of copper at 10.30 cents a pound,[43] which paid for deepening and concreting the shaft, preparatory to exploration of this and other ore masses. Time was running out for the company, however. Copper receded to seven cents a pound, and mining companies faced the worse time in the history of American mining. Before delving into that story, however, let's backtrack and look at some of Lemuel's other business ventures. At the same time we'll see how events of the 1920s affected the Shattuck family.

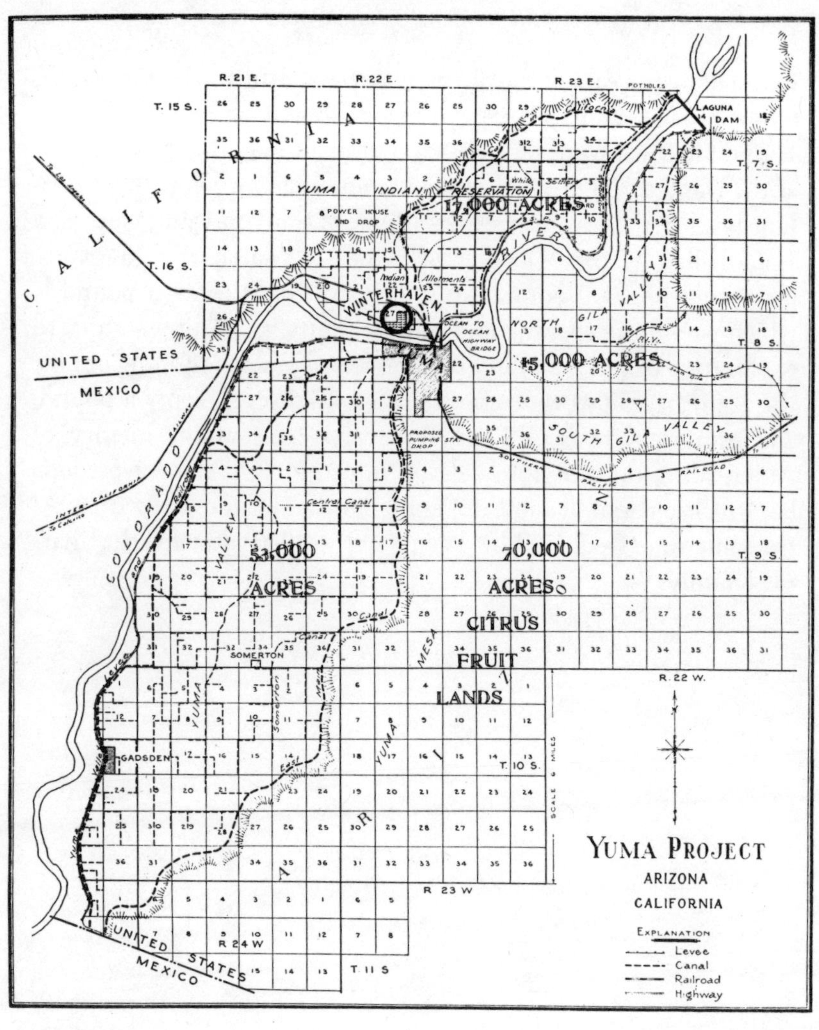

R. 21 E.
R. 22 E.
R. 23 E.
T. 15 S.
CALIFORNIA
YUMA INDIAN RESERVATION
17,000 ACRES
POWER HOUSE AND DROP
T. 16 S.
WINTERHAVEN
Indian Allotments
YUMA
OCEAN TO OCEAN HIGHWAY BRIDGE
NORTH GILA VALLEY
POTHOLES
LAGUNA DAM
T. 7 S.
T. 8 S.
UNITED STATES
MEXICO
15,000 ACRES
PROPOSED PUMPING DROP
SOUTH GILA VALLEY
SOUTHERN PACIFIC
RAILROAD
Central Canal
COLORADO
VALLEY
T. 9 S.
53,000 ACRES
70,000 ACRES
CITRUS FRUIT LANDS
Canal
MESA
YUMA
R. 22 W.
SOMERTON
YUMA EAST
T. 10 S.
GADSDEN
R. 23 W.
R. 24 W.
UNITED STATES
MEXICO
T. 11 S.
SCALE 6 MILES
N
YUMA PROJECT
ARIZONA
CALIFORNIA
EXPLANATION
Levee
Canal
Railroad
Highway

TEN

THE WINTERHAVEN
COMMERCIAL COMPANY

FOR YEARS copper, cattle, and cotton were the bulwarks of Arizona's economy. By virtue of his ties to mining and ranching, Lemuel Shattuck figured large in developing the first two industries. His leap into cotton was daring, prompted by a little altruism and a search for investments, stimulated by mining associates. Both Byron and Martin Pattison, promoters of the Shattuck-Arizona and Denn-Arizona Copper companies, eyed the 40,000-acre Blythe Ranch as a future cotton farm. In 1911 they attempted to lure Lemuel Shattuck into a joint purchase of acreage, and the venture was attractive to him.[1] Land about Blythe was fertile and well-watered. Long, hot summers and mild winters were conducive not only for cotton, but for melons, citrus, and alfalfa. At forty dollars or less an acre the price was right. Always the astute businessman, Lemuel saw one drawback — no railroad — and turned down the proposition.

While he sidestepped the Pattisons' venture, Lemuel did not close his mind to potential in the Colorado River Valley, and watched with interest the progress of the five million dollar Laguna Reclamation Project near Yuma, as well as developments in Imperial Valley. At the same time, he began a long correspondence with the U.S. Department of Agriculture and agricultural experiment stations in California and Arizona relative to water resources, irrigation, and soil conditions of the Imperial and Yuma valleys. He even inquired as to whether water from the Salton Sea could be used to irrigate crops.[2]

The Southern Pacific Railroad spanned the Colorado River and crossed Yuma Valley in 1877-78, but agriculture literally did not bloom until completion of the Laguna Dam in 1907 and the Yuma Reclamation Project three years later. A complex of syphons, floodgates, pumps, and canals rendered 145,000 acres on both sides of the Colorado River susceptible to irrigation. The Valley became a land of opportunity when the Bureau of Interior opened the Yuma Indian reservation to farming, by offering 173 farms of forty acres each for as little as $66 per acre.[3]

With both a railroad and irrigation, the Yuma Valley was attractive in 1910-11. A keen-eyed miner with the caution of a banker, Lemuel weighed investment possibilities for several years. As a cattleman he knew land. "Anyone desiring to buy land around Yuma had better take a careful look at the ground before making a trade," he wrote to Martin Pattison. "There is some raw Yuma land that can be leveled and put in cultivation for about $30 to $40 per acre, while other tracts cannot be properly leveled and put in cultivation for $1000 per acre. It all depends on the nature and contour of the ground."[4]

In August 1914 he finally moved, purchasing two tracts of land south of Yuma: 160 acres near Gadsden and 320 acres near Somerton.[5] Admitting he "did not know much about farming," Lemuel placed the smaller parcel into the care of his son Henry, who was majoring in agriculture at the University of Arizona. Wiley Fitzgerald, a trusted mining man who had supervised Shattuck mineral properties in Sonora, cleared mesquite and leveled the Somerton land. Irrigation ditches were dug, linking both tracts to the canals of the Reclamation Project, and the farms brought into production with sorghum and alfalfa.

Shattuck did not have to be a farmer or a visionary to know that a federal irrigation project watering 128,000 acres on the Arizona side of the Colorado River and 17,000 acres across the river would turn the valley into a garden. Here was more potential than just Yuma could absorb. Shattuck grasped the opportunity. Within six months of his move into the valley, he was eyeing 160 acres adjacent to the Yuma Indian reservation on the California side of the river. Situated within 600 feet of the Southern Pacific tracks, this level piece of bottom land appeared ideal for a townsite and commercial center. Collaborating with C. Evan Lucas and A. E. Deyo, real estate brokers, and L. L. Odle, the Yuma Indian agent, Shattuck invested $39,000 to purchase the land from J. R. Kerr.

Shattuck and his associates filed incorporation papers for the Winterhaven Townsite Company on December 31, 1915, at Imperial, California. Directors of the corporation were Lemuel C. Shattuck (president), Bisbee lawyer C. T. Knapp, C. Evan Lucas (business manager), A. E. Deyo (treasurer), and L. L. Odle. Initially capitalized at $40,000 — $39,000 of which was subscribed by Shattuck — the company would build a winter resort for West Coast vacationers and a commercial center for farmers on both sides of the Colorado River. Forty acres were cleared by Yuma Indians under L. L. Odle's supervision, surveyed, and divided into a residential and a commercial district. Lots were advertised for $250 each, payable at $10 a month. Included with each lot were 250 shares of stock in the Winterhaven Improvement Company, a public utility that would furnish electricity and community services. All stock sales, escrows, and mortgages would be handled through Shattuck's Miners and Merchants Bank in Bisbee.

In autumn the Winterhaven Commercial Company was organized, consisting of a general merchandise store and a lumber company. A bank and a hotel with a liquor license were envisioned; the latter a drawing card for thirsty farmers from dry Arizona. Attachment of financier L. C. Shattuck to the venture instantly drew subscribers from Bisbee and Douglas, and Lucas's skilled advertising pitch of "invest dimes today — realize dollars tomorrow," lured investors from San Diego to Globe. Within a year 125 individuals, including Joseph Muheim, J. T. Hood, Arthur Houle, and Dr. Nelson Bledsoe, had purchased lots at Winterhaven — in all $119,829 worth of real estate. Indeed, it looked as if Shattuck and his associates would "have a town that will make people sit up and take notice before the public is hardly aware of it."[6] Equally exciting to Lemuel were thousands of acres of land about Winterhaven then being cleared for small farms of from 40 to 160 acres.

A letter shattered Lemuel's Winterhaven dreams on April 20, 1916. Yuma lawyer W. F. Timmons, and his partner W. O. Harris, had purchased through C. Evan Lucas several prime lots in the Winterhaven Townsite. Before receiving deeds to the property, the attorneys authorized Lucas to resell the property for no less than $1,000, Lucas to retain as commission all proceeds above that amount. Lucas sold the lots to Joseph Muheim for $1,100, but turned over only $255. Unable to pry the balance out of the broker, Timmons turned to Shattuck: "I am unable to secure from Mr. Lucas anything satisfactory and am convinced of his intention to defraud me . . . out of the balance purchase price of these lots, to-wit $745. I can ill afford delay in the matter and will greatly appreciate anything you may do to the end that I may receive an early settlement of what is due me." Shattuck did not want to get imbroiled in this matter, but knew a charge of fraud would hurt development of the Winterhaven Townsite. "I will have a talk with Mr. Lucas," Lemuel promised Timmons.[7]

Shattuck was an accepting person who believed most people were innately honest. When an individual proved otherwise, he was gruff and matter-of-fact, inclined to seek punishment to fit the offense. In Lucas's case, the banker did not just "talk." He told the salesman to pay Timmons what was owed. Adding that he would tolerate no shady dealings in the sale of Winterhaven property, Shattuck forced Lucas to resign as manager of the townsite. Unfortunately, the man stayed on as real estate salesman. Dumb as he was crooked, Lucas never paid the lawyers, and Timmons filed a charge of embezzlement, resulting in Lucas's arrest on July 19 as he boarded a train for Los Angeles.

The Yuma papers played up the news for all its worth, painting Winterhaven as a den of crooks. Shattuck was deluged with letters inquiring into the financial stability of the townsite. A number of buyers of Winterhaven property demanded their money back, and a larger number stopped monthly payments. "This is a hard knock to Winterhaven to have our former manager and present salesman arrested for embezzlement," the embarrassed banker wrote. "I do not know whether he is figuring on getting out of Winterhaven but one thing I do know . . ., if he does not change his ways in regard to getting us in bad I will surely figure on getting out of Winterhaven myself."[8]

Lucas raised bail, and departed for central and southeastern Arizona, pushing Winterhaven real estate as if nothing had happened. In two weeks he turned in sales receipts for twenty-seven lots. Because Lucas promised anything to make a sale, trouble dogged his trail. He sold a lot to Benjamin and George McNelly, Bisbee auto mechanics, and pledged to erect a building large enough to handle automobile repair work. Flat broke, Lucas offered a third of the capital necessary to promote a gas station.

A letter from the McNellys again drew Shattuck into the fray. "I think it is a shame for Lucas to promise to build a building for these boys when he knew that he could not build it," Shattuck wrote Deyo. "Things like this are sure to get us in bad with the people who have already bought lots, as these boys are sure to express their opinion around Warren and Bisbee about the way they have been treated and I do not blame them either for doing so."[9]

This time Shattuck was the one who contemplated filing criminal charges against Lucas. The salesman was flighty, never staying in one place long enough to permit serving of a complaint. He was reported to be promoting a mining venture in the Globe-Miami area, selling stock in Mesa. Finally, he and his wife decamped to Canada, leaving behind a string of creditors and bilked investors.

The taint of fraud displeased the California Corporation Commission, and in November 1916 it refused to grant a charter to the Winterhaven Improvement Company. Without a charter the Improvement Company was prohibited from furnishing electricity and other services to the townsite, forcing Shattuck to refund money collected on the purchase of eighty-five lots. Lucas's fraudulent dealings killed Winterhaven as a community, but the

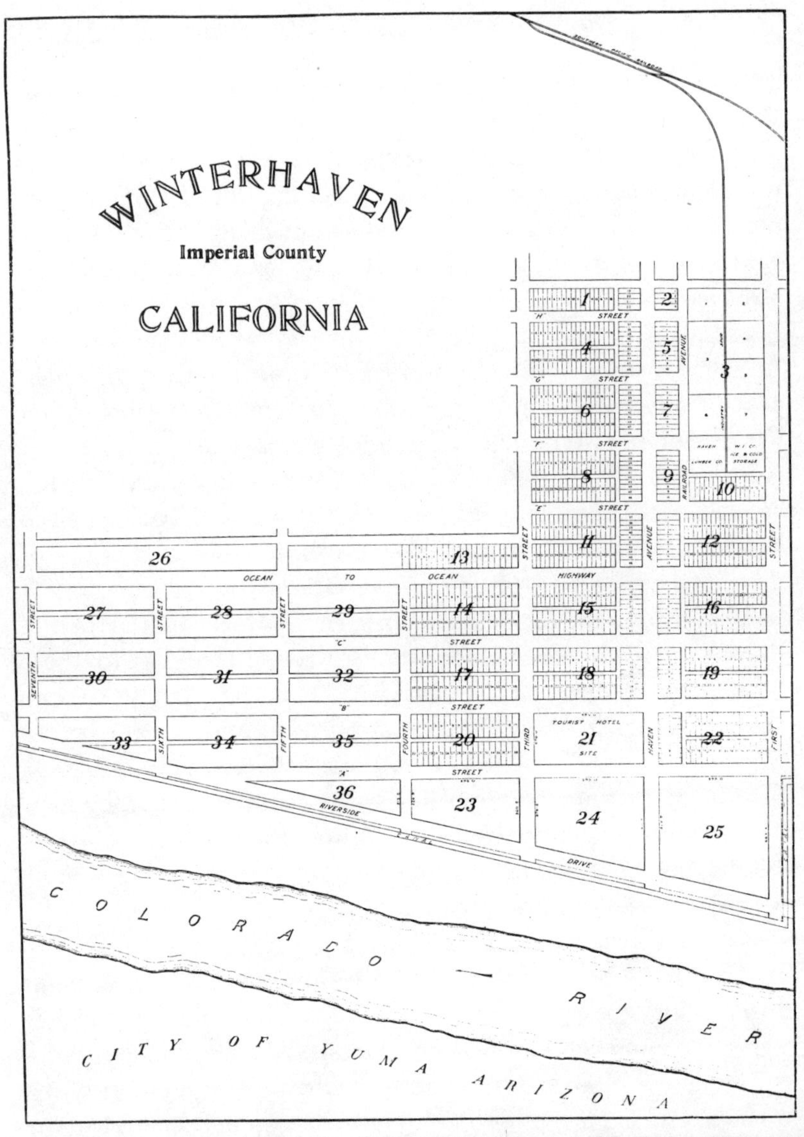

INVEST DIMES TODAY----REALIZE DOLLARS TOMORROW

dream of a commercial and farming center still burned in Shattuck's imagination.

Despite failure of the townsite, the Winterhaven Commercial Company, as a mercantile establishment, was in an enviable position. It was the only supplier of farm implements and construction material on the California side of the Colorado, and freight charges to Winterhaven were less than to Yuma, allowing for effective competition against E. F. Sanguinetti, Yuma merchant and the valley's leading citizen. Unfortunately, Lucas's resignation and disappearance left A. E. Deyo with not only management of what was left of the townsite, but supervision of the general store and lumberyard, the latter two enterprises badly undercapitalized. Although he had advanced money to the Commercial Company to cover startup costs, Shattuck did not want to bail A. E. Deyo out of a financial jam. To do so would mean advancing the company additional funds, securing them with a chattel mortgage. In short, he would be assuming responsibility for another business. Strange as it may sound, Shattuck was saved that headache by C. Evan Lucas.

During his rambling in Globe, Lucas encountered J. P. Glass, who had just sold the Globe-Miami Lumber Company. Glass had capital and was looking for an investment—music to the ears of any real estate promoter. Lucas spent the hot summer of 1916 driving Glass all over the Yuma Valley, from north to south. Crops that year were excellent; the prices farmers were getting pointed to considerable construction in the fall. In fact, Deyo had already signed up $3,390 worth of building. Glass was convinced that "no other place offers the opportunities that Winterhaven did."

Wanting to be free of the lumber business, Deyo signed it over to Glass, "as it stood," for $9,500, which covered the actual cost plus handling charges for all stock, equipment, and sheds. On August 29 Glass put $500 down and agreed to assume the business on October 1.[10] With $40,000 in his pocket from the sale of his Globe lumber business, Glass put in a full line of lumber and went after the construction trade. To draw farmers to Winterhaven, he increased the stock of agricultural implements and wire. And he proposed to go fifty-fifty on a cotton gin. Here was an offer that Shattuck could not turn down for a number of reasons.

Through his research, Shattuck was aware as early as 1910 that cotton could be grown in the Yuma and Imperial valleys. He was especially interested in H. J. Webber's development, near Calexico, of a strain of Egyptian extra-long staple cotton known as Jannovich. As a student of agriculture, Henry Shattuck further perked his father's interest with reports of improvements of this strain by T. H. Kearney of the Arizona Experiment Farm at Yuma. When Henry planted in 1915 a small acreage at the Gadsden farm to a new breed of Jannovich known as Yuma extra-long staple, Lemuel observed firsthand the superior attributes of hybrid cotton. The fiber of Yuma cotton was one and one-half inches long; strong and lustrous, better than the short-staple Durango grown in the area under sponsorship of E. F. Sanguinetti. Added to that, Yuma cotton produced no less than two bales per acre. It was Henry's opinion that Yuma cotton would shortly be the only variety grown in the Valley. And Lemuel concurred.[11]

Despite problems besetting the Townsite Company, Lemuel matched Glass's offer of $9,500. A. E. Deyo and Susie Odle each contributed $100, enough to give them a voice in company affairs as board members, and the Winterhaven Commercial Company was launched "to do a general mercantile and banking business, to build and operate cotton gins and deal in cotton and other products, to build and operate cottonseed mills and deal in the products of the same; to buy, lease, sell, rent farms and farm lands; to generate and sell electricity and other power for domestic and power consumption private or public; to manufacture, sell and deal in ice and other products. . . ."[12] The commitment was made. Shattuck had stepped from copper to cotton.

Deyo and Glass spent November and December of 1916 convincing farmers on the California side of the Colorado River Valley to plant Yuma cotton. It was a task made easy by the fact that this type of cotton was already being grown on several farms across the river, including Henry's, and in limited quantity in the Salt River Valley and at Parker. By year's end seventy-two farmers had pledged to plant 1,068 acres to extra-long staple cotton, provided that a gin would be available to process the crop by August 1, 1917.[13]

In his methodical manner, Lemuel took up the almost superhuman task of putting together a ginning company. With Glass and Deyo handling everyday business technicalities at Winterhaven, Shattuck scrutinized and passed on every detail relative to establishing the ginning operation: from purchasing equipment, arranging its transportation, to buying seed. As a banker who knew the pitfalls of financing farmers, he formulated the terms whereby the Company would underwrite the cotton crop.

The contract initially drawn up by Deyo specified the crop be financed with loans secured by mortgages, and that farmers pay for ginning.[14] This Lemuel opposed. "It is dangerous," Shattuck counseled Deyo, "to loan money to a bunch of cotton farmers to make a crop and pick and haul it to the gin, and attend to all the details. And believe me, there will be some work to attend to the details of these loans."[15] The best method of finance was always the simplest in Shattuck's way of thinking—to loan against ginned cotton in a public warehouse. To gain farmers' support for the cotton enterprise, Deyo and Glass had committed the Commercial Company to financing a crop. Shattuck had no choice but to agree, insisting the notes, secured by chattel mortgage against tangible property, be endorsed and assigned to the Miners and Merchants Bank, which in turn would loan the Winterhaven Commercial Company money at six percent interest. The Commercial Company had to attend to collection of all notes.[16]

Shattuck passed judgment on equipping the project, just as he did on financial matters. To maintain an adequate supply of water for the townsite and its enterprises, a 25,000-gallon hemispherical bottom steel tank was purchased and erected atop a sixty-foot tower. Lemuel compared power rates of several utility companies in Arizona and California, and finally negotiated a contract with the Yuma Light and Power Company, personally guaranteeing the expense of stringing a transmission line from its substation to Winterhaven.

An improved self-feeding, twelve-stand, long staple cotton gin was purchased from the Continental Gin Company of Dallas, Texas, for $5,748. As this was a cash purchase, Lemuel furnished the money.[17] The gin was installed on a concrete foundation within a shed, under supervision of E. G. Attaway of the Mesa Egyptian Cotton Exchange. Also erected were storage bins for raw and processed cotton, and seed. The laying of a spur line, 600 feet long, was negotiated with the Southern Pacific Railroad to connect the gin with the railroad's main line.

Throughout July a crew of twelve carpenters labored in 110-degree heat, erecting the ginning complex on First Street in the northeast corner of the industrial section of Winterhaven. Fifteen hundred acres of extra-long staple cotton were maturing about Winterhaven. Across the river 2,500 to 3,000 acres had also been planted to the same type of cotton. Whether or not this crop came to Winterhaven depended on installation of a gin at Yuma or Somerton. As yet there had been no movement to erect a gin capable of processing extra-long staple cotton, although E. F. Sanguinetti, who had several short-staple gins, was toying with the idea.

Building was almost completed by end of July, but Yuma Light and Power Company had yet to install lines to Winterhaven. With only sixty days remaining before ginning commenced, there was very real fear that no electricity would be available to drive the machinery. By August 7 Shattuck could wait no longer and tersely ordered R. M. Moore, manager of the utility company, to "get busy and build the power line at once."[18] In the event the utility company did not deliver on time, Shattuck took the precaution of securing prices for diesel and gasoline power plants. Fortunately, he did not have to resort to installation of backup equipment, for Yuma Light and Power began stringing wire, and on September 9, three days before start of ginning, the transmission lines were connected to transformers at Winterheaven.

No sooner had Shattuck solved one problem than another arose. With only a tenth of the labor needed to harvest the crop, farmers were bidding against one another for fieldhands, driving up their

The Winterhaven Commercial Company in 1917. This photo was reproduced from company stationery in the Fathauer collection.

crop's labor costs by seven to eight cents a pound. Shattuck solved the problem by securing 500 Mexican nationals from the American Beet Sugar Company, thus assuring the cost of picking at no more than four cents a pound. As farmers were unable to pay for this labor, the Commercial Company advanced $3,000, secured by Lemuel, adding the expense to ginning charges. [19]

Picking began in mid-October with ample labor. The cotton came in fast, filling the bins, and more was stored at various farms, awaiting transportation to Winterhaven. Ginning commenced early in the morning of November 11, and by nightfall thirteen bales stood in the yard. A night crew was hired to keep the gin running around the clock. [20] Shattuck, Glass, and Deyo had pulled it off — they had brought in Yuma Valley's first commercial crop of extra-long staple cotton, much to the consternation of E. F. Sanguinetti.

Entrance of the Winterhaven Commercial Company into cotton production was thought to be well-timed. Upon commencement of World War I Britain placed a virtual embargo on Egyptian cotton, denying American industry, then gearing for war, an essential component to munitions, airplanes, barrage balloons, and uniform fabric. Like copper and other vital commodities, cotton soared in price. Winterhaven farmers could have easily obtained seventy cents a pound for their 1917 crop. They chose instead to hang onto the cotton in hope of getting a better price in 1918.

With a large inventory of baled cotton and hopes of good future prices, Glass and Deyo began signing up every cotton grower in the Yuma Valley. With the view of garnering the Yuma cotton crop, they created the Winterhaven Egyptian Cotton Exchange in early 1918. Consisting of about ninety growers, this cooperative would be capitalized at $50,000, $30,000 to be raised by assessing each grower five dollars an acre for cooperative ownership of the Winterhaven gin. A like assessment would be made the following year to erect a gin at Somerton.[21]

While the Winterhaven Commercial Company had taken a giant step in Arizona agriculture, financial trouble threatened to halt its momentum. Deyo and Glass had loaned farmers over $300,000. Supposedly secured by mortgages on crops and ginned cotton, this money was obtained from Shattuck's financial institution. The Miners and Merchants Bank, however, faced a crisis of its own, brought on in large measure by the war. To insure solvency of the national banking system, the newly created Federal Reserve Board insisted its member banks place twenty percent of total assets into treasury certificates: about a million dollars, in the case of the Miners and Merchants Bank. If that were not enough, each Liberty Loan and Red Cross drive that swept the Warren District shrank deposits by $250,000 to $300,000, forcing Shattuck to halt credit to Winterhaven.[22]

If the Winterhaven Commercial Company were to survive it had to sell the hundreds of bales of cotton stored in its yards, and collect the notes extended to farmers. F. C. Ketter, a hard-nosed forty-year-old accountant from Riverside, California, was hired to turn both cotton and notes into cash, a task made difficult by wartime priorities.

An audit of the notes revealed an alarming lack of sense in managing the Winterhaven Commercial Company. "I find that the business as far as loans on chattel mortgages is concerned, is in a very bad shape," Ketter informed Shattuck. "Some mortgages . . .

The ginning facility of the Winterhaven Egyptian Cotton. This photo is also reproduced from company stationery in the Fathauer collection.

were made without any security whatever, mostly for clearing land on Indian leases and some for building purposes. As these mortgages call for a growing crop as security, and in several cases there being no crop, it seems to me that these loans were made without any judgment or proper investigation of the applications." This was not what a banker wanted to hear, and Shattuck summoned Ketter and the notes to Bisbee.

Obviously the Commercial Company was in trouble, and if something was not done, the Miners and Merchants Bank would come out the loser. Shattuck froze all loans and sent Ketter back with instructions to "investigate every loan, and if necessary, foreclose unless proper additional security is furnished." And Shattuck came down hard on those responsible for making the loans.

"If my instructions are not carried out as to cleaning up the loans," he warned Glass and Deyo, "our bank will make no further loans on either growing crop or ginned cotton. If necessary I will send an attorney to Winterhaven to help Mr. Ketter clean up this mess, and when the thing is cleaned up I will quit."[23]

Shattuck fumed on one hand and was conciliatory on the other. "If, however, these loans are taken care of . . . we will advance farmers that have a growing crop of cotton reasonable amounts to make their crops, but they must be reasonable amounts, and the crop must be inspected before the loan is made."[24]

Because he controlled the purse strings, Shattuck's words were complied with. By March 1918 the $115,000 overdraft at the Miners and Merchants Bank had been reduced by $85,000, and Shattuck relaxed credit restrictions, allowing the Commercial Company to loan an additional $300,000. This time loans were calculated on the basis of $25 an acre, the money secured by chattel mortgages. Contrary to past practice, the money was doled out a little bit at a time: five dollars per acre payable when the crop showed "a good stand." Additional sums were then issued for labor and purchases of supplies. In this way, Shattuck ensured the Miners and Merchants Bank would be "protected at all times."[25]

To insure survival of the farmers and collection of notes, Shattuck and Ketter had to sell cotton. Originally they planned to dispose of the bales through the Winterhaven Egyptian Cotton Exchange. Assessment of acreage fees, however, caused the organization to disintegrate into a bunch of squabbling farmers, incapable of fielding any representation among commodity buyers. The alternative was to place cotton with Sanguinetti's Arizona Products Company, something Shattuck would never do. He chose instead to bypass middlemen and attempt to sell cotton directly to New England mills.

Shattuck directed Ketter to option the cotton and load it onto railroad cars for shipment East. Accordingly, four carloads of cotton were consigned to Springfield, Massachusetts, and on June 14 Ketter, carrying bills of lading for 117 bales (the first two cars), boarded the train for New England.

To dispel the impression among mill purchasing agents that Ketter "was out to sell cotton at any price," Shattuck provided him with a letter of introduction from the Miners and Merchants Bank, which not only gave him authority to broker Winterhaven cotton, but portrayed the Bank as the "financial bulwark" of the Winterhaven Commercial Company. To keep secret Ketter's movements in the East, the Bank would send bills of lading on subsequent shipments to banks designated by Ketter, and handle collection of sales through National Shawmut Bank of Boston, the

Miners and Merchants Bank's only correspondent in New England.

Upon arrival in Springfield, Ketter consigned bills of lading for the first two carloads of cotton to the National Bank of Commerce in Providence, and the next two cars' paperwork to the Springfield National Bank. Warehousing the cotton with the Bay State Storage and Warehouse Company, he then set about contacting buyers. Throughout July Ketter called on mills, without success.

Many New England mills were struck by textile workers demanding higher wages to keep abreast of wartime inflation. Uncertainty regarding the length of the war and persistence of government contracts was evident everywhere. Buyers for twenty-one of the twenty-six mills Ketter called on told him they would not be in the market for cotton until fall. Others merely refused to see him or put him off. "Consumers are stagnant . . . waiting to see what will happen," Ketter reported to Shattuck.[26]

Ketter was not one to sit idly by waiting for a decision from purchasing agents. He coaxed a petition from cotton growers in Yuma and Imperial counties taking the government to task for encouraging the planting of $1.5 million worth of extra-long staple cotton — on which banks had loaned half a million dollars — for which there was no market. Thus armed he headed for Washington, D.C., hoping to sell Winterhaven cotton directly to the government for balloon and airplane fabric.

He had no luck in that respect, but found New England mills somewhat responsive when he returned from Washington, D.C., in August. At least purchasing agents were willing to talk, requesting samples of Winterhaven cotton that kept Ketter busy pulling cotton for testing. When the staple of his cotton proved longer and stronger than the best grades of Sea Island and Egyptian cotton, inquiries came in as to availability of 500- and 600-bale lots. E. A. Shaw & Company of Boston, the most reliable cotton broker, gave a "vague indication" that they could dispose of 1,200 to 1,300 bales.

Price slowly crept into discussions when purchasing agents informed Ketter that by purchasing in lot, mills could get Egyptian cotton laid down in Springfield for 46 cents a pound. This was discouraging news for Shattuck, who hoped to sell cotton in the neighborhood of 70 cents a pound. Knowing that only Pima cotton, a new hybrid, would be raised after 1918, Shattuck insisted that Ketter make every attempt to sell their Yuma cotton, even if they had to take less than 70 cents.[27] Anticipating no better than 56 cents a pound, Shattuck thought it best to ship East all cotton stored in the Winterhaven yards. At least it would be immediately available should a sale be perfected.[28] At end of August Ketter was summoned back to Winterhaven by Shattuck to coordinate the shipping of cotton. He stepped off the train a sick man, suffering from laryngitis and an acute bronchial infection contracted in the warm, damp cotton storage sheds. For all his efforts, Ketter was greeted with only surliness and suspicion by Yuma Valley farmers.

Cotton farmers had placed great faith in Shattuck and Ketter selling their cotton in the East for 65 to 70 cents a pound. Their failure to do so provided ammunition for the Winterhaven Commerical Company's greatest enemy, E. F. Sanguinetti and his Arizona Products Company.

Known as "Yuma's First Citizen," the "Merchant Prince," and "Yuma's Greatest Benefactor," Eugene Francis Sanguinetti was in many respects similar to Lemuel Shattuck. Born in Coulterville, California, of immigrant Italian parents on May 16, 1867, Eugene arrived in Arizona about the same time Shattuck did, taking a job in John Gandolfo's mercantile store at Yuma in 1883. By 1890 the bright, aggressive youth was a partner in the enterprise, and owned it by the turn of the century. A dynamo, the short, dark, and muscular Sanguinetti expanded the mercantile business to mining camps up the Colorado River, grubstaked prospectors, established restaurants, organized the Yuma Electric and Water Company, and even had a string of undertaking parlors. The mainstay of his fortune, however, was an ice making plant that furnished refrigeration for the Southern Pacific Railroad. And he had the Coca Cola franchise in Yuma; it was not Anheuser Busch, but it made money. He did stock the finest wines and hard liquor at his store.[29]

As the largest farmer and buyer of produce in the Valley, as well as a dealer in agricultural implements and hardware, Sanguinetti viewed Shattuck's entrance on the Yuma scene with suspicion and contempt. "What does that saloon keeper know about cotton," the Italian once voiced to Indian agent L. L. Odle. His view of Shattuck as a mining camp spoiler was a projection, and he attempted to undermine Shattuck's progress at every turn. He struck at Winterhaven by spreading the rumor that Ketter could easily have sold cotton for 70 cents, but upon Shattuck's orders chose not to, and that the cotton instead had been shipped to New York, where George R. McFadden & Company advanced money on it, allowing the Miners and Merchants Bank to collect interest on farmers' notes.

Sanguinetti further aroused farmers by claiming he had sold a carload of cotton at Galveston for 65 cents a pound, and had cotton at Chicopee Falls, Massachusetts, that he expected to sell for 70 cents. Sanguinetti's report infuriated Yuma Valley farmers to the extent that some demanded the Winterhaven Commercial Company release their cotton so they could either sell it themselves or place it with other brokers. Actually, Sanguinetti had no better luck than Shattuck in getting a satisfactory price for cotton. He too was forced to warehouse cotton by the strike of Chicopee Falls mills. Had not Lemuel Shattuck quickly responded, Sanguinetti's ploy would undoubtedly have succeeded.

Lemuel Shattuck did not sink to the level of rumor-mongering. He singled out James H. Maxey, the most irate and outspoken farmer seeking release of his cotton, and spoke his mind:

I have been informed that a number of my good friends (?) [question mark in the original] have been poisoning the minds of a number of the long staple cotton growers in the Yuma Valley, and that the rumor is growing that the Winterhaven Commercial Company could have sold the cotton for 70 cents in Chicopee Falls, Mass., but that the company did not wish to sell it for the reason that they wanted to collect the interest on the notes; also that they had shipped this cotton to New York and had borrowed money from George R. McFadden & Company. . . . I wish to state a few FACTS to you.

. . . Neither the Winterhaven Commercial Company nor the Miners and Merchants Bank have borrowed any money in the East on this cotton.

We are holding the bag for you fellows and loaning you money at 6%, and I guess you know that we could loan every dollar that we have out at 8%, so we are not making any money by loaning to the farmers at 6%. It is only a matter of accommodation because it seems that everybody got in bad for the reason that they did not sell their cotton last season. If the farmers at Yuma had been willing to sell their cotton for 70 cents in the early part of last season — or last fall — things would have been altogether different.

I assure you that no one is more interested in the quick sale of this cotton than I am, as we have about $300,000 tied up which we could use to a whole lot better advantage.[30]

Although Shattuck's quick response halted Sanguinetti's highjinks, the banker nevertheless knew that if Winterhaven cotton was not sold growers would come at the company with "hammers to knock us in every way." He hurried the shipping of cotton to such an extent that by mid-September 1,207 bales were warehoused in Massachusetts. Ketter headed East with Shattuck's instructions ringing in his mind: "It is essential that we sell the cotton . . . it will turn a whole lot of money loose and our bank will be in a better position to finance this year's crop."

Ketter spent sixty days conferring with purchasing agents, and drawing samples for comparison by mill operators. Winterhaven cotton was well received, most operators stating that it had arrived at the right time, for Sea Island cotton was off forty-six percent and selling for 70–75 cents, and the entire crop of Egyptian extra-long staple cotton was expected to be off forty percent. Conceded as good as the best grades of Sea Island, Winterhaven cotton was expected to fetch at least 73 cents a pound. Being spot cotton, it had a good chance of exceeding that price.[31] Placing bills of lading with Massachusetts banks, and samples of cotton with purchasers, Ketter returned to California to await commencement of buying in January.

The first inquiries regarding price of Winterhaven cotton came in on January 14, 1919, to which Ketter responded by quoting no less than 73 cents for Yuma variety, and 78 for what Pima variety the company had. Buyers bristled at these prices, stating the best grades of Sea Island could be had for 52 to 54 cents delivered to New Bedford or Fall River. At Ketter's prices there were no takers of Winterhaven cotton. The November armistice ending World War I left the New England fabric industry stuck between wartime contracts and peacetime aspirations, hesitant to purchase in large lot. Then too, there was every indication that low priced foreign cotton would soon be available.

Other than consigning twenty-five bales with the Pierce Spinning Company of New Bedford, "to get them started," Ketter failed to sell any cotton.[32] Shattuck momentarily rescued the Winterhaven Commercial Company by disposing of 700 tons of cotton seed at sixty-one dollars per ton to the Ballinger Cotton Oil Company and the Winters Cotton Oil Company, both of Texas.[33] The $42,700 from this sale allowed the company to carry on its ginning operation, which broke a Yuma Valley record of twenty-one bales a day on January 24, 1919.

On expectations of unloading cotton for at least 70 cents a pound, the Commercial Company had advanced farmers as much as fifty cents per pound, and the Miners and Merchants Bank held $300,000 worth of notes. With the Eastern market frozen by indecision there was little chance the cotton could be sold for enough to cover loans to farmers, let alone insurance, freight, and interest on notes. It was a double-bind situation for Shattuck. He wanted to protect his Winterhaven interests and help Yuma Valley farmers on one hand, but as a banker he was obligated to ensure the solvency of the Miners and Merchants Bank. No matter which way he turned someone would surely be hurt. He sought the least painful course.

Late in January 1919 Shattuck went to Yuma and put the facts before the farmers. In a series of meetings with growers, he explained that collapse of the American commodity market made it imperative to sell Yuma cotton at market price, and sell quickly. If that could not be done, as a banker who must protect his financial institution, he had no alternative but to call the loans due on the first (1917-18) Yuma cotton crop. In the meantime, he told the farmers, the bank was willing to rewrite some notes, taking "some of the future cotton crop as collateral." It was piling debt upon debt, but most farmers agreed.[34]

As treasurer of the Commercial Company, the job of collecting and rewriting notes would fall to Ketter. He had returned from the East, however, broken in health, hemorrhaging from bronchial infection contracted during the 1918 sales trip. Shattuck could not pry him out of convalescence in Riverside, California. In desperation, Lemuel sent his assistant cashier, I. F. Burgess, to Winterhaven to supervise the transactions.

Burgess found the Winterhaven Commercial Company a financial "mess," besieged by creditors. Someone in the Lumber Company had been ordering merchandise on Commercial Company stationery. On top of that, A. E. Deyo, manager of the general store, had twice mortgaged his milo maize crop, moving his indebtedness onto company books, and Ketter had not returned bills of lading for much of the cotton sent East.

If that were not enough, the fifty-two bales of cotton consigned to the Pierce Spinning Company sold for only 48 cents a pound — and the drastic drop in price was not due to economic factors. The Goodyear Tire and Rubber Company had eliminated demand for spot cotton by putting out contracts for automobile tire yarn and cloth with numerous Eastern mills with the understanding that mills would buy Pima cotton through the Goodyear Company.[35] Little wonder that Burgess reported the "darn farmers not coming in as fast as I hoped they would"[36] to pay and renew their notes. He doggedly kept after them, however, and by end of March had reduced the Commercial Company's $93,000 overdraft at the Miners and Merchants Bank to $35,000.

In the meantime, Ketter's health worsened. On April 23, his physician informed Shattuck that the cotton broker was running a 104 degree temperature, could not speak, and was "slightly irrational and rambling on subjects other than cotton and Winterhaven."[37] Still able to read and respond to telegrams, Ketter was lucid enough to negotiate cotton sales when the market suddenly came alive. On April 26, he wired Shattuck "several hundred bales Yuma and Pima sold today." By mid-May he had sold 963 bales of Yuma and Pima cotton at an average price of 51.84 cents, or $242,151.[38] Another 500 bales, including some of the 1919 cotton crop, was sold on May 21. This was the last cotton sold by Ketter. Still hemorrhaging and suffering from pleurisy, he entered a sanitarium the next day, forcing Shattuck to place cotton sales in the hands of A. M. Clemens of the Stewart Brothers Cotton Company, a New Orleans brokerage company with offices in New England.[39]

Cotton sales continued: out of a total of 1,791 bales shipped East, the Commercial Company disposed of 1,568 bales in June; the last 223 bales were sold in August. The Commercial Company's account with the Miners and Merchants Bank stood at $285,972 by mid-summer, and Shattuck went to Yuma to settle with the farmers.

For three weeks he endured the 108 degree heat of the Winterhaven store, writing checks to farmers not in debt to the company, and going over notes of heavily mortgaged growers. The latter complained about the average of 51 cents a pound obtained for their cotton. When Shattuck explained that freight, insurance, interest, and other fees would erode the amount even more they turned suspicious. Some farmers were able to cover the shortfall of their notes; most could not, and the Winterhaven Commercial Company came up $93,000 short. Shattuck returned to Bisbee to convey the news of the deficit to directors of the Miners and Merchants Bank, leaving Burgess in Yuma to manage collection of delinquent notes and foreclosures.

Because Ketter had not paid freight on cotton as expected, pocketing instead more than $6,000, it took Burgess weeks of labor in stifling heat to settle accounts. Finally, on July 16 he wrote Shattuck: "I absolutely will not stay any longer in this unGODly hot place. Can't sleep and can't eat. Have been down here now five months, away from my home, wife, and friends, and that is rather too much. I haven't turned bolsheviki or anything like that, but I've got to get out of here."

"I am sorry that the heat is getting your goat," Shattuck sympathized. "I did not think it would take as long as it has to get out these cotton settlements, but just as soon as the settlements are finished I think you had better take a vacation for a while." In the meantime, Shattuck directed Burgess to "take a shot of bootleg once in a while to keep your courage up."[40]

Although Ketter made good on some freight bills, his attempted fraud infuriated Shattuck. Not only was it faith-shattering, it held up the settlement with farmers. Tracing bills of lading through Eastern banks and brokerage houses incurred extra expense. Lemuel was scrupulous and expected the same from his subordinates. Even though Ketter was near death, the banker called a meeting of the board of directors of the Winterhaven Commercial Company, fired Ketter and directed Yuma lawyer I. T. Colman to prepare a criminal complaint.

Ketter's actions were not the only aggravations besetting Lemuel Shattuck and the Commercial Company. With ending of World War I, Egypt began dumping cotton on the American market, causing Arizona farmers to turn to other crops. Only 500 acres in the Yuma Valley were planted to cotton at end of 1919. The most serious problem facing Shattuck, however, was the Winterhaven indebtedness carried by the Miners and Merchants Bank — an indebtedness bank examiners howled about.

The bank carried three notes: one for $50,000, another for $40,000, and a smaller indebtedness of $2,640. All were personally guaranteed by Shattuck. It was lack of any interest payment for eighteen months on the first two notes that drew the attention of bank examiners. On top of these notes, the Bank was carrying $62,000 in farmers' notes. As the Commercial Company was in the process of collecting the latter notes, the Bank was willing to "give a little more time on them." It was a different matter with the three notes aggregating $92,640. These notes, Bank directors ordered, "must be taken up on or before January 1, 1920."[41]

This notification, a legal formality, spelled death for the Winterhaven Commercial Company. To cut his losses, Shattuck had no choice but to begin dismantling the enterprise. Upon conclusion of settlements in August 1919, he began casting about for buyers of the gin. When he did not receive any concrete offers for the machinery, he leased it in September to W. T. Bunton, one of few remaining cotton growers in the valley, who also purchased, for $4,000, some of the central addition of the Winterhaven Townsite.[42] In November the greater portion of the townsite was acquired by Arnett Pawley, of Bard, for $7,000, who immediately leveled the land in preparation for planting alfalfa. On December 20, 1919, Shattuck, J. P. Glass, I. F. Burgess, and Susie Odle, comprising a quorum of directors, voted to liquidate the company.[43]

Shattuck paid the Winterhaven indebtedness to the Miners and Merchants Bank, took back from the company a note amounting to $118,140, and proceeded to sell remaining assets. He disposed of fourteen lots, six houses, and miscellaneous structures to C. H. Kent of Yuma, who also optioned to buy the cotton gin for $20,000. When Kent's option lapsed in spring of 1920, it was offered to a series of parties including E. F. Sanguinetti, A. D. Putnam representing the Goodyear Company, and B. B. McCall. It was finally sold on May 12 for $15,000 to McCall Cotton Company. In December 1922 Lemuel purchased for $15,000 — the amount to be applied against the company's indebtedness to him — all remaining assets of the Winterhaven Commercial Company, including fourteen lots in the townsite, notes and accounts receivables, trust deeds, machinery of two farms, and livestock consisting of ten mules, eight horses, two colts.[44]

Following Shattuck's acquisition of these assets, the Commercial Company was disincorporated, thus ending a venture founded in 1915 with the highest of hopes. At that time it never entered Lemuel's mind the enterprise might end so disastrously. "I thought that I would at least make a stand-off when I went in with

Lucas and Deyo in the townsite and cotton business, and that I would be doing considerable good for the country there by boosting the long staple cotton game," he wrote to L. L. Odle. A combination of crooked dealing, inept management, and a paralyzed cotton market following the war soured Shattuck to the point that by end of 1919 he was "desirous of closing out all my interests in Winterhaven, take my losses and forget about the whole affair."[45]

ELEVEN

THE NOT SO ROARING TWENTIES

THE armistice of November 1918 brought joy to Bisbee on one hand and hardships on the other. At close of World War I the price of copper stood at twenty-two cents a pound. When government support of the red metal was removed early in 1919 the price slumped to eighteen cents, forcing mining companies to cut production. In February 1919 the Calumet & Arizona Company closed its Oliver and Cole shafts, the least productive of its mines. The Copper Queen halved its production and cut wages by a tenth. Commodity prices slumped further, copper reaching fourteen cents a pound in December 1920.

Businesses in the community responded to deflation by cutting prices. J. C. Penney announced "Reconstruction Prices" based on replacement cost, and Frankenberg and Newman reduced prices at their Fair Store by fifty percent.[1] A general and steady reduction in prices of food and merchandise was equally apparent in other stores.[2] Roominghouses announced a fifty percent reduction in rent. Perceiving a worsening economy, Bisbee's bankers urged their customers to patronize local businesses and save money. The Miners and Merchants Bank reflected concern in this advertising ditty:

> Where are the dimes of yesterday?
> Nickels and dimes,

> Nickels and dimes,
> How freely I spent you
> And thoughtlessly lent you
> I've wanted you back again, scores of times,
> Nickels and dimes![3]

On December 22 the bottom dropped out of the stock market, and three months later copper retreated below twelve cents, the line between profit and loss. Mining companies ceased production, the smelters at Douglas were shut down, and two-thirds of Bisbee's 5,000 miners were unemployed.

Miners drifted off in search of work, some to the Ranger oil fields in Texas, others to Signal Hill, California, and the fields around Bakersfield. Bisbee was a town of empty houses. As in 1907–09 Bisbee's commercial circles felt the pinch, a pinch made more painful by Prohibition. Effects of the previous depression had been softened by the town's wide-open liquor and gambling trade. Not this time. Seekers of good times who still had cash spent their money in Naco and Agua Prieta, while those without resources lounged on the porch and steps of the library, sat in soft drink parlors, or aimlessly wandered the streets. "The hotel and rooming-house owners are not doing any business, as nearly all of the people who are in Bisbee are married and have families. I do not know of a roominghouse in Bisbee that is paying their rent," observed Lemuel Shattuck.[4] "Bisbee has never been as dull in its history as it is now," Shattuck commented to another longtime friend.[5]

Withdrawals instead of deposits were the order of the day in all banks. The Bank of Bisbee, backed by Phelps Dodge, easily weathered the decline, as did the Miners and Merchants Bank, with over 3.6 million dollars in assets. Both banks sharply reduced credit. The Citizens Bank, Bisbee's third largest bank, was closed by bank examiners in August 1923.[6] There were not enough employed miners in the southern end of the district to support the Bank of Lowell, and Shattuck and Michael Cunningham closed it in late 1923.

It took several years to reduce the worldwide glut of copper. In 1922 the price of metal began to creep upward, and mining companies responded by bringing mines and mills back into production. By September 1923 red metal stood at 13½ cents a pound and Bisbee's mines were once again in operation, including the much depleted Shattuck-Arizona. As explained in Chapter 9, increased metal prices enabled Lemuel to push development of the Denn, setting the stage for merger of the two companies.

Resumption of mining brought solace to Shattuck. During the course of five years he had suffered both financial and personal loss. The $92,000 Winterhaven loss was nothing compared to the disappearance of Henry. Although the stern, outwardly stoic Shattuck spoke little of his son, some hope the boy would appear still burned within him as late as 1924. On April 8 of that year he directed Warner, who was investigating a low grade copper property near Tepic, to "size up the Americans around there and look them over. Maybe Henry might be around there, but I do not think so."[7]

The guilt of sending Henry to Mexico ate at Belle, perhaps compounding a diabetic condition that had been developing for a number of years. Following the Shattucks' return from Central America in 1921, her health improved somewhat, then slumped again. Lemuel thought another trip south of the border would benefit Isabella, and in 1923 he announced that as soon as school was out for the summer, the Shattucks would tour Mexico. Isabel, Dorothy, and Spencer were elated. Isabella had a dressmaker make several dresses for the girls, dotted swiss and voile outfits that would be cool.

Early in June the family drove to Nogales, Mexico, and boarded a night train. They awoke the next morning to see the sun rise over Guaymas, the blue ocean meeting the dry desert hills, under spreading sun rays. Fishermen were preparing to set out. The train stopped and a small troop of soldiers, with camp followers, were put aboard two flatcars: one behind the engine and one at the end of the train. They were there to protect the train from attack by revolutionists.

It was a slow journey, for the train halted every few hours to take on small bundles of wood to fuel the engine. At each stop the soldiers jumped off and "El Capitan" lined the sorry troop up for drill. One had the pants of the uniform, another the jacket, another the cap. Only their leader had a full uniform. Often the men broke ranks to ask El Capitan for a match to light a cigarette. Belle doubted these soldiers could offer any resistance should rebels attack the train. During the stops, vendors appeared out of the countryside with tortillas, tamales, live iguanas, and armadillo roasted on the half shell. Lemuel tasted the latter, but not Belle or the children.

The train rolled on, passing countless small one-room, doorless and windowless adobes having only three walls. The fourth side was open to the elements. Scantily dressed peons sat inside on hard packed dirt floors. Lemuel told the girls they were poor Indians who would be happier and fare better in their native hills.

Arriving in Mazatlan, Lemuel guided the family to the Belair, a modern, well-built hotel constructed by a German who had immigrated to Mexico after World War I, along with others of his nationality. They had started new lives, organized enterprises such as the Belaire, and built and operated breweries that gave Mexico its *cerveza*, as well as beer gardens akin to those of their former country. They married Mexican women and produced industrious offspring.

Although the Belair Hotel was on the seashore and received a pleasant offshore breeze, mosquito netting hung over every bed, and prior to retiring the family had to chase insects out of their rooms. Warner joined the family, and the next day Lemuel and the boys hired a fisherman and his boat. Belle, Dorothy, and Isabel lunched on the patio. Appearance of a large boa constrictor, however, disrupted their relaxed meal; Isabella grabbed her children and ushered them upstairs to their rooms. When Lemuel returned, she scolded him and told him about the snake. He merely smiled and explained that such serpents were kept in hotels, warehouses, stores and other places of business to cope with mice and rats. He asked the hotel manager to lock up the boa.

The manager apologized to Belle the next day, and explained that the snake had eliminated all rodents, and now had to be fed. A few days before a *mozo* had failed to lock its cage and the boa had consumed a small fox terrier, which shocked Isabella. The next morning the children were invited to view the reptile in its cage, the *mozo* took the snake out, draping it over his shoulders so everyone could have a close look and touch it.

A noisy ruckus on the floor where the Shattucks were staying disrupted their sleep every night, causing Belle to complain loudly — to no avail. General Rodriguez was staying in the hotel with his retinue and they were celebrating a victory over the rebels. At lunch the following day, a pretty young red-haired woman, accompanied by a dark, dour *duenna*, set down at a nearby table. The young lady kept looking at Belle, as if wanting to talk. Lemuel advised his wife not to respond, for Rodriguez was an important man in Sonora, and the lady was his wife, an American from Chicago. After the Shattucks returned to Bisbee, the *Daily Review* reported the young woman had committed suicide.

The Shattucks took the train to see Russell Grenfell, Belle's brother, in Culiacan, Sinaloa, a beautiful town of streets lined with tropical trees bearing large white flowers, *flores de muerte*, which Hispanics picked and used for funerals. Lemuel installed the family in the Hotel Rasales in the colonial part of town, a block or two from Russell's home. It was a charming old two-storied building surrounding a patio on three sides. A wall on the fourth side, with massive wooden doors, shut off the outside world. Keys to the rooms were huge; the doors large and carved; the rooms spacious with high ceilings and handsome armoires.

Each morning the Shattucks were awakened by the sing-song of street vendors, selling fresh pineapple juice, mangoes, bananas, parrots and *hielos*, iced drinks, from push carts. Church bells rang and rang, urging people to attend Mass. Dogs barked day and night. Belle accumulated the family's dirty laundry, and asked Lemuel to find a laundress. He did, and when the clothes were returned full of holes, Belle was furious. The clothes had been carried to a stream and the soiled spots beaten with stones. Lemuel laughed at local sanitation, and reminded the family they were visitors in a country not as modern as the United States. The always adaptable Shattuck advised his children to be gracious and appreciate Mexican culture and ways of living. Belle and the girls could accept almost everything about Mexico, except the beggars who took up stations at street corners and on church steps. Their revulsion was not the least bit altered by Lemuel's explanation that begging was a respected, often lucrative profession in Mexico.

Lemuel demonstrated courage in taking his family south of the border, into the midst of revolution. American mines had been seized, and in some cases personnel slain. Several mining engineers had been taken from a northbound train and executed. Mormon colonies in Sonora and Chihuahua had been overrun, and refugees were streaming into the United States, among them the Williams family, whose son, Jack, became governor of Arizona years later. Twenty thousand Mexicans sought sanctuary across the border. Despite this violence, Lemuel traveled in Mexico at least twice a year to inspect mining sites and visit Warner. Wherever he was, he knew leaders on both sides of the revolution.

The trip into Mexico improved Belle's health. Her recovery did not last, however. Diabetes again debilitated her. Dr. Nelson Bledsoe, chief surgeon of the C&A Company, treated her to the best of his ability, and told Lemuel that injections of insulin, a new and little known drug, might stabilize her condition. It was at Bledsoe's urging that Shattuck contacted the Rockefellow Institute for Medical Research of New York in spring of 1923. Lemuel was directed to Ely Lilly & Company, the manufacturer of insulin,[8] and a supply of the drug was procured, which Bledsoe administered daily.

In fall of 1924 Isabella was stricken with appendicitis. Lemuel rushed her to the Copper Queen Hospital on October 22, and she was operated upon by another doctor, Bledsoe being out of town. Weakened by diabetes, Belle's immune system could not stave off infection, and she became critically ill. When it appeared she would not survive, Lemuel broke the news to Isabel.

He entered the room and sat beside his eldest daughter while she was doing homework, reading Spanish. He took her hand, and in a distraught voice related how Belle had been an exceptional wife and mother, always loving. He told his daughter they had started with very little, working together all these years. Her intelligence and judgment he respected and sought for all ventures, small or gigantic. She had never failed him; he could not have succeeded without her. Belle was his helpmate.

Isabella Grenfell Shattuck died on Thursday, October 24, 1924. While not publicized in either Bisbee or Douglas newspapers, her funeral three days later in Bisbee's Episcopal Church drew friends and business associates from throughout Arizona, including Governor W. P. Hunt and Senator Henry F. Ashurst. She was buried in the Evergreen Cemetery beside her two sons.[9] Notes of condolence poured in from mining men, cattlemen, saloonists, gamblers, frontiersmen, and oldtimers who loved and respected the Shattucks. The Arizona Bankers' Association adopted the following resolution, and it was forwarded by Morris Goldwater, secretary of the Association:

"WHEREAS, the Supreme Master had taken from our midst the beloved wife of our friend and associate, Mr. L. C. Shattuck, President of the Miners and Merchants Bank of Bisbee;

BE IT THEREFORE RESOLVED, that the members of this Association extend to Mr. Shattuck their sympathy in his hour of bereavement and that the sorrow and sadness which visits the home of a member banker is partially born and shared by the members of this Association at large.[10]

Belle's death crippled Lemuel for some time. He could not attend to daily activities. Appointments were canceled. "Too much upset," he called off a trip to Paradise to assess a lead-silver property owned by George A. Walker.[11] Although he had three children in Bisbee, and Mark and his wife, Frances, stayed on after the funeral, Lemuel experienced a sense of loss he had never known. Loneliness paralyzed him. He developed indigestion, did not eat, and lost weight. "Things are not the same as they used to be around the house," he confided to O. M. King, an old mining man. "It is pretty tough to lose a partner that you have been with for over thirty-five years. Anyhow these things must happen to every one."[12] A fundamental strength saw him through difficult situations, and he expressed that faith to Spencer: "Things come pretty tough sometimes, but the only thing for us to do is to do the best we can."[13]

Mark and Frances lived with Lemuel for nearly a month, and when they departed, seventy-three-year-old Joe Martinelli, who had managed the gambling concession at the Shattuck saloon, moved in. Belle would never have permitted a gambler in the house, but things were different now. The old man from Switzerland was the best possible therapy. He looked after Lemuel, saw to it that he ate, soothed loneliness. Martinelli even watered Belle's garden, and fed the dog and cat. Other friends came around. Arthur "Dutch" Koch and Leo Jeleniewski appeared nightly, and the four men played cribbage until early morning. Shattuck gained weight and resumed his daily activities. His wry sense of humor returned. "Joe Martinelli and myself are batching at the house," he wrote son Spencer. "He is not a very good cook but I am a dandy and we have just what we want. He gets up and makes the fire and I get up and help cook. Inside of twenty minutes we have the finest breakfast you ever saw. At night we sometimes have a mulligan with everything in it around the house except the woodpile. Sometimes we have macaroni and the way we fix that up it has most everything in it too. We get along fine and my stomach is a whole lot better. I can eat most anything now within reason."[14]

Isabella's death left Lemuel with four sons (he had not given up on Henry), and two daughters to worry about. Eighteen-year-old Spencer, in his second year at Stanford University, was not much of a problem, so long as he kept out of the clutches of girls. Warner was a cause for concern, however. Relaxation of penalties for draft evasion brought Warner to San Francisco shortly before Isabella died. While Spencer reported him "happier than he ever saw him,"[15] Warner found California's coastal climate "too damn cold," and began preparing to depart for Mexico. At Spencer's suggestion, Lemuel offered Warner a job supervising exploration at the Estancia Mine in Quila, Sinaloa, at $200 a month.[16] Warner took the job, and Lemuel turned his attention to his girls.

What to do with his daughters was a dilemma. Isabel, age fifteen, would graduate from high school in January 1925, and Dorothy, barely thirteen, was just beginning high school. With his daily schedule at the bank, frequent trips to Mexico, and dashes East for board meetings of the mining companies, Lemuel was in no position to ride herd on two daughters.

He decided to get them out of Bisbee, and enroll them — as he had his sons — in the finest schools in the San Francisco area. In that way, they would be close to Spencer, and could visit one another on weekends. He enrolled Isabel at Mills College in Oakland, and Dorothy at the Castillejo Girls' School in Palo Alto. Mark, who had moved to California, assumed the task of watching over the children.

Always a dapper dresser, Mark's first concern was that Isabel and Dorothy each have a proper wardrobe. "Both of the children have to be outfitted from one end to the other," he advised his father, "and their things cost equally as much as things for my wife or any other grown person . . . When they go to school it is necessary that they have nice things like other girls in school. This is going to take some money to fix them up. . . ."

When it came to his children, Lemuel was never tightfisted. He would fix them up in proper style, and he provided Mark with $1,265 to appropriately dress the girls. Shattuck gave all his children a liberal allowance of $200 per month, but set one condition. Each child had to write to him at least once a month. Failure to do so would result in forfeiture of the monthly allowance. This law came as no surprise. The Shattuck home had always been regulated.[17]

Mark's role as surrogate parent was compounded by the reappearance of Warner in fall of 1925. He had worked the Estancia Mine for over six months, sinking two shafts. When only ledge matter turned up, he folded the operation and returned to San Francisco to be close to his brothers and sisters. His periodic drinking bouts and wanderlust, however, plagued Mark, and worried

Mark Shattuck at age twenty-four.
(Fathauer collection)

Lemuel. As with his sisters, Mark looked after his older brother, reporting on his condition regularly to Lemuel.

Dorothy adjusted quickly to the Castillejo School, and, according to Mark, "made some real nice friends. The regular hours have done lots for Dorothy. All the pimples have left her face and she looks fine." The college environment weighed heavily on Isabel. She fretted about algebra and other studies. "The work seems very hard after just coming from high school," she wrote her father. "Golly, we sure don't have much to eat." She lost thirteen pounds, although the school doctor pronounced her in good health.

Lemuel sent her an elementary algebra book, and was not at all sympathetic over the lost weight. "I guess you can stand that thirteen pound loss of flesh all right, as you were too fat anyhow and I do not think you will starve."

Gruffness masked Lemuel's love. "I note that you had your physical examination and that the doctor pronounced you in good shape. You will have to have an intelligence test also, and I hope that he will pronounce you normal and not too much of a monkey. Hope that you . . . will get a chance once in a while to show off some of those clothes I bought you as there is no use having them unless other people can see them."[18]

Isabel may have found school difficult, but she was never a recluse. She grabbed the first opportunity to show off her clothes during a night on the town, which drew an approving and revealing comment from Lemuel. "Sister is getting to be a sport," he wrote Spencer. "She is going to attend Grand Opera with Mark and Frances in San Francisco. I do not go much on Grand Opera myself. Would rather see a good vaudeville show any time. The last time that I attended Grand Opera in Chicago, four of us left before the play was finished because we were laughing and did not appreciate same, so we left before the management threw us out and went to the Chicago Opera house to a good vaudeville."[19]

Mark fussed over the girls like an old hen. "Isabel is complaining that she has more work than she can handle at school and says algebra is awful. . . . Isabel is about all in from studying . . ., her eyes are about half closed. When she was over this last weekend she was so tired she could hardly move around and said she did not want to fail. A girl friend described her as a nervous wreck.

"Lem, she is sure a bag of nerves," Mark confided to Lemuel, and suggested Isabel slow down her studies, something Isabel would never do.[20] She told her brother to quit clucking. She would study all night if necessary. Isabel conquered algebra, finished her first semester, and returned to Bisbee for summer vacation along with Spencer and Dorothy.

Lemuel always kept close tabs on his children, and was aware Spencer had fallen in love with a girl from New York, not at all unexpected for an eighteen-year-old youth. But when the girl turned out to be six years older than Spencer, and her parents announced an engagement without informing the Shattucks, Mark concluded they "were a plain bunch of gold diggers."[21] Lemuel respected his children's privacy and did not immediately pry into Spencer's personal life, figuring his son would relate the details in due time. What Spencer revealed, during a trip to the family's ancestral home in Erie, Pennslyvania, did not please Lemuel. The object of his affection was Helene Lewis. She was indeed six years older than Spencer, and they had eloped to Reno.

Lemuel thoughtfully smoked his pipe a while, and finally said, "I will send money to your wife and ask her to come to Bisbee. But when she arrives, I want an hour alone with her."

When the woman arrived, Shattuck invited her into his office, closed the door, and studied her before saying: "You are five or six years older than my son. I could have the marriage annulled, for my son is still a boy and you are a grown woman. I will give you a chance to make the marriage work. But if Spencer sows a few wild oats later, please don't come to me. I expect him to graduate from Stanford. I will pay his tuition, give him an automobile, and provide a monthly allowance for you to live on."

The summer of 1926 was eventful for the Shattucks. Joe Martinelli departed the house on Brewery Gulch, going to California to visit a brother, making room not only for Spencer and his wife, but for Dorothy and Isabel, the latter on vacation from school. Aurelia Henry Reinhardt, the president of Mills College, also arrived in

Bisbee to visit families of six other students attending her school. One of America's few women college presidents, Aurelia was interested in mining, and Lemuel conducted her through the smelters at Douglas, showing her the surface facilities of the Denn Mine, with its newly installed diesel engines. He held on to her while she leaned over the shaft to see and feel the hot steam pouring from the depths. She questioned Shattuck about hardrock mining and the process of smelting copper, and Lemuel was attentive, giving precise answers. Later Aurelia asked Isabel what college or university Lemuel had graduated from. When Isabel replied he had never attended college, the erudite lady was astonished and said that her father had a tremendous fund of knowledge, and was better educated than many college graduates. Lemuel, in turn, admired Aurelia Reinhardt, and commented on her great intellectual curiosity. He always appreciated an intelligent woman.

It was because of another intelligent woman that Lemuel and Spencer journeyed to Erie, Pennsylvania. Lemuel was attracted to Mary O. Wilson, a comely thirty-five-year-old school teacher from Warrensberg, Missouri, who taught in the Bisbee schools, and more recently in Douglas. They saw one another regularly and by summer of 1926 had laid marriage plans. Not wanting to settle a bride into Isabella's house on Brewery Gulch, Lemuel began searching for a residence in Warren. Three houses struck Mary's fancy: the Notman house, the Ovens home, and the house of John Treu. After considerable dickering, Lemuel purchased the latter residence, at 200 Vista, for $9,770,[22] and furnished it with refurbished heirlooms from the Shattuck home in Erie, which he and Spencer had secured on their trip east. On the morning of November 25, 1926, the Reverend Eason Cross married Lemuel and Mary O., as she was called, in a quiet ceremony in the rectory of St. John's Episcopal Church in Bisbee.[23]

Shortly thereafter, Lemuel and Mary O. left on an extended honeymoon trip through western Europe, taking Isabel and Dorothy. At least one trunk of clothing was packed for each of them, and loaded aboard a luxurious transatlantic liner. Everyone

Mary O. Wilson about the time of her marriage to Lemuel Shattuck.   (Fathauer collection)

dressed in formal wear for dinner each night, but that was of short duration, for Dorothy and Mary O. took to their beds with seasickness. Like diehard seamen, Lemuel and Isabel paced the deck, and indulged in thought-provoking discussions, touching on moral issues of the day.

Shattuck talked about bootleggers and the numerous speakeasies in the United States, and the availability of dangerous liquor. He knew flappers and college students considered it sophisticated to smoke and drink. His teenage daughters must learn to use judgment about alcohol and cigarettes. He did not want them sneaking behind his back to drink or smoke. Instead, Lemuel allowed Isabel and Dorothy to smoke or have a cocktail or a glass of wine, if used with control and discretion.

The liner docked near the Firth of Forth in Scotland, and the Shattucks cruised the British Isles, crossed the Channel, and took a train from Le Havre to Paris, where they were met by an American Express agent who immediately taxied them to a hotel. He advised them to remain in their rooms, for a large mob was in the streets protesting the execution of Sacco and Vanzetti — something that Lemuel had to witness. He left the hotel and went to the site of the protest, which had turned into a riot, in time to see mounted gendarmes, swinging billy clubs, disperse the shouting crowd. As always, Shattuck had to witness the action, be it in America, Mexico, or France.

Europe was beautiful in the late 1920s, for the different countries had distinct identities and cultures, no congested traffic, and few factories. The Riviera was unspoiled. Germany was striving to restore herself after defeat. Holland, with its many canals and flowers, was exquisite. Italy stood in sharp contrast to her northern neighbors, and conditions there reminded the Shattucks of countries south of the United States' border. In France, Lemuel insisted the family eat horse meat served as a main course. There, cattle raised on a limited land base produced only milk and cheese. Calves went to market to provide delicious veal. There was no beef. In Rome the Shattucks met two American girls who had run out of funds. They were ill and almost desperate. Lemuel gave them enough money to return home and saw them on their way.

The Shattucks returned to Bisbee, and Isabel and Dorothy went back to school in the fall. A few months later, at two o'clock on the morning of October 13, 1928, Mary O. gave birth to a seven-and-a-half pound baby boy, "the result of a big all Shattuck event," Lemuel announced to the family by telegram. Had the baby arrived two hours earlier, Lemuel joked, he would have been christened Christopher Columbus.[24] He was named Lemuel Coover. But at sixty-two years of age, fatherhood for Lemuel had lost much of its luster.

At the annual family Christmas gathering in Bisbee that year, Lemuel confided to his children that he did not feel the same thrill over a birth that he did when younger. Life is full of challenges, choices, problems, happy and sad times, and he could not help but wonder what fate would deal this child. There was sadness about Lemuel. "It is easier to die than live, he said, for most people slip into a coma and pass away." He had seen many die.

Lemuel had a young wife and an infant son; both would outlive him by many years — enough to make any man feel his vulnerability. Heavy purchases of insurance in 1927–1928 indicate that their welfare weighed heavily on him. Shattuck was also at an age where he was witnessing the demise of close friends. On May 7, 1927, Joe Martinelli left Bisbee, "looking just like a young fellow," to visit a brother in California. A week later he was dead.[25] Shattuck had been close to his head gambler since the founding of the saloon in the 1890s. The deaths of Belle and Martinelli were the sunset of a bygone day, made all the more certain by the settlement of his deceased wife's estate. Since Arizona is a community property state, the $1.6 million estate — the largest probated up to that time — was divided among the children, the inheritances set aside as trusts administered by Shattuck.

Mary O. was a skillful homemaker and gracious hostess, but soon found the Warren home too small for entertaining. In 1929 Lemuel purchased the McGregor House, across the park from the Treu residence. The home was open to all Shattuck children, who felt protective of "Little Lem," as they lovingly called their half-brother. Nevertheless, an era had ended for the five children of Lemuel's first marriage. Only nostalgic memories remained, as Isabel recalled to her father: "I used to love you to tell me stories at night. The old house in Erie seemed so fascinating to me. . . . I often thought how happy we were at the old house in Brewery Gulch. It was so cozy evenings and you and Mama used to talk and read the paper together."[26]

Distribution of Belle's estate in May 1927 amply provided for the five Shattuck children, allowing the boys to enter the world of business and high finance. It assured the girls the best educations, and freedom to pursue artistic and cultural endeavors. Upon graduation from the Castillejo School in 1928, Dorothy matriculated at the Oberlin Conservatory of Music, and Isabel transferred to Northwestern University at Evanston, Illinois, majoring in economics. Within a year, Dorothy become disenchanted with stodgy Oberlin, and joined her older sister at Northwestern.

Spencer graduated with honors from Stanford University in July 1927, returned to Bisbee, and assumed the duties of teller at the Miners and Merchants Bank. Mark procured a franchise from the Pathex Company to sell motion picture projectors throughout Florida, and moved to St. Petersburg in early 1926. His business collapsed, however, when Pathex sent two salesmen into his territory to undercut prices. Mark returned to Los Angeles and assumed the managership of the Shattuck family assets in the Yuma Valley: the farms and Winterhaven property. His inheritance enabled him to speculate in the stock market, and he bought heavily into the petroleum industry, some $40,000 on margin, and on tips from his father, moved in and out of Shattuck-Denn stock, which netted him considerable profit. As the mining company prospered in the waning years of the 1920s, he invested more in its stock until he had 6,300 shares by summer of 1929.[27]

Warner also played the stock market, but not to the extent that Mark did. Like his father, he knew mining stocks were a dangerous investment even in the best of times, and he sold most of the Shattuck-Denn stock received in his inheritance. The unpretentious

Warner lived frugally and disappeared for lengthy periods of time, causing Mark to wring his hands over his brother's "toots." In true Shattuck fashion, the younger brother invariably tracked Warner down, sobered him up, and put him back on his feet.

No doubt the four youngest Shattuck children considered the 1920s "roaring," in the sense that it was a period of coming of age, self-discovery, and freedom from financial worry. To Lemuel, however, the decade whimpered more than it roared. Throughout most of the period it was all he could do to maintain his empire.

The first pillar of the Shattuck empire was mining, and the 1920s was not a good time for that industry for several reasons. The copper glut following World War I collapsed the metals market in 1922–23, forcing closure of mines, and revolution in Mexico drove American companies out of that country. Shattuck and Byron M. Pattison had holdings at Cananea known as the Juanita Mining Company. Because of turmoil in that district they could do no more than maintain yearly taxes, in hope of someday returning to develop the property. By end of the decade, however, they had ceased paying taxes, and their claims reverted to the Mexican government.

As seen earlier, the Shattuck Mine was depleted by 1925. Except for ground containing limited lead reserves, the mine was turned over to leasees who extracted ore from pillars left by previous mining operations. Although Lemuel shifted his focus to developing the Denn Mine, that operation did not begin to pay until 1928-29, when commercial quantities of ore were shipped that netted the company about $100,000 a month. In the meantime, Lemuel had contributed $150,000 of his personal fortune to bring the mine into operation.[28] As he had not received any mining dividends since 1921, this outlay of cash imposed some hardship.[29]

The economic climate of the 1920s also buffeted the Miners and Merchants Bank, the second pillar of the Shattuck empire. At close of World War I, Lemuel affiliated the Bank with the Federal Reserve System. Since the Federal Reserve required member banks to carry larger cash reserves, the pool of cash from which the bank made loans shrank. Regardless of that handicap, Lemuel expanded deposits to make up the shortfall, and generously financed southeastern Arizona commerce and industry. By 1921 the Miners and Merchants Bank was the largest bank in Cochise County, with assets of 3.5 million dollars.

As a bank serving a one-industry town, the Miners and Merchants Bank was vulnerable to the cyclical ups-and-downs of copper, and Lemuel endeavored to broaden the institution's customer base, often going far afield to do so. Besides catering to local mining, the Bank financed small business in Cochise County, dairy herds in the Salt River Valley, farmers at Yuma and Parker, and cattlemen throughout Arizona, as well as town and county construction projects, the latter underwritten by bonds that netted no more than four to six percent. The Miners and Merchants Bank even had a hand in financing the lucrative liquor trade along the border throughout the Prohibition era.

Lemuel, who had never severed his ties with oldtime saloonists, befriended Frank O. Mackey, an El Pasoan who held a large interest in the Gadsden Hotel, and who had worked years before in the Shattuck saloon. Besides carrying the first mortgage on the Douglas hostelry, Shattuck frequently advanced Mackey money between 1922 and 1926 to develop the Kentucky Distillery Company of Juarez, Mexico, which made fine whiskey. As a banker with a background in the liquor trade, Lemuel went a step further than just rendering financial aid. He did not hesitate to counsel Mackey on business matters: "If I were you I would raise the price of the whiskey you have in stock and make nothing but grain alcohol for quick delivery to the rectifiers in Juarez." Shattuck also put Mackey in touch with reputable Sonoran wholesalers, among the latter A. L. Dominguez of the Monterey Brewing Company, who did "a large business along the border towns and bears a good reputation in handling everything straight without fixing things." With aid from the Miners and Merchants Bank, by 1926 the Kentucky Distillery was operating "full blast, . . . running just as smooth as water through a greased pipe," making rye and bourbon.[30]

Financing men engaged in the border liquor trade was profitable, and there were few defaults, most of the individuals being long-

Lemuel C. Shattuck in the mid-1920s. (Fathauer collection)

time saloonists who valued a "square" reputation. Cattlemen were of the same mold, and so long as good economic times reigned, Shattuck went out of his way to help them. Unfortunately, deflation of the early 1920s was accompanied by a circle of droughts that crippled the livestock industry of southeastern Arizona, New Mexico, and West Texas. For six years out of ten, rainfall was below normal, compelling some stockmen to drive their animals into Sonora, where grass was more plentiful. High freight rates precluded supplemental feed, and maintenance charges were too expensive to permit fattening, resulting in premature marketing and financial loss.

The disastrous combination of deflation and drought is reflected in Shattuck's rejection of loan applications made to the Bank. In January 1923 the loan committee rejected a loan request for $250,000 submitted by E. A. Tovrea to refinance his Arizona Packing Company in Phoenix.[31] When Willis, son of Enoch "Aus" Shattuck, requested a loan of $3,000 to see the Shattuck Ranch of Ashland, Kansas, through a difficult time, Lemuel reluctantly had to turn him down, saying "our bank is not in a position to make any loans as we are having a hard time to carry the cattle and business men in our vicinity." Forced to dip into his personal resources to cover the $92,000 Winterhaven loss, as well as to finance development of the Denn Mine, Lemuel sadly wrote, "I am sorry to say that I have no money personally to loan you and have considerable obligations myself to be taken care of. . . ."[32]

Lemuel would undoubtedly have helped Willis Shattuck if he could, for he was always generous when relatives approached him with requests for sound investments. His correspondence during this period, however, indicates he had his hands full just keeping his fortune intact and the bank on even keel. He did endeavor to maintain credit lines established by long-standing customers. Such was the case with the Riggs Cattle Company of Willcox, Arizona, which had a revolving credit line of $50,000.

Late in December 1921 banker-cattleman William M. Riggs and his brother, John C., informed Shattuck that they were dissolving the Riggs Cattle Company and dividing its land among stockholders, who would still retain a percentage of the herd based on stock holdings. Termed "an adjustment for the future," the move was calculated to eliminate squabbling over inheritance and estate at a later date. Shattuck acknowledged the wisdom of the decision, renewed a $5,000 balance on the original note, and added that "if you wish larger amounts after the company is dissolved, you can give us a mortgage on your and your brother's part of the land."[33]

Evidence that Shattuck was attempting to maintain his financial liquidity is seen in his turning down the purchase of two of Arizona's largest ranches. When aging Douglas banker-cattleman, B. A. Packard, offered the Turkey Track Ranch, a giant 100,000 acre spread lying along the border, "having as good a title as can be made on this continent," and stocked with 4,000 cattle, Shattuck tersely responded. "He did not care to increase his cattle holdings."[34] In May the following year, the Perrin Trust Committee offered Shattuck the Babocomari Grant, which the Miners and Merchants Bank had regularly financed, and had appraised at $150,000 several years before. Shattuck, however, turned down the Perrin deal just as he had Packard's offer. He suggested, however, that the Hearst Estate, which owned the Boquillas Land and Cattle Company on the San Pedro River, might be interested in acquiring the grant. Apparently it wasn't, for Shattuck would be offered the Babocomari Ranch six years later.[35]

There were several reasons for Shattuck declining these offers. First, it was all he could do to keep his two small ranches going. The drought of 1924–25 had destroyed the cattle industry in southeastern Arizona. Wells and tanks at the Bar Boot Ranch had dried up. A gulch three-quarters of a mile above the ranch was dry for the first time in forty years. Shattuck and Meadows succeeded to restoring that water source by scrapping out the accumulated sand, but a well drilled at the O.K. Ranch never hit water, forcing Meadows to water cattle at the Winkler Well and at a spring on the forest reservation. In desperation, Shattuck shipped steers to pasture at the Shattuck Ranch in Ashland, Kansas, and sent 350 O.K. cows and a 150 Bar Boot cows to Fiant, California, thirty miles northeast of Fresno, on the San Joaquin River. The O.K. cattle were later sold for thirty-five dollars a head, including calves. Being "old skates," the Bar Boot cattle could not be sold.[36]

The paramount reason for Shattuck declining to be land rich was that he was bringing the Denn Mine into production, a task that ate up tens of thousands of dollars of his savings. A lesser reason was that the company's board of directors were casting about for additional mining property. Archibald M. Chisholm held 500 acres in Whitney and Tisdale townships of Ontario, Canada, close to the big Hollinger, MacIntyre, and Schumaker gold mines. In summer of 1927 Chisholm and Thomas Bardon, Jr., hauled Shattuck to Canada to inspect the property. When no apparent mineralization turned up, the directors of the Shattuck-Denn focused their attention on the hills above Hayden Junction, about seventy miles northeast of Tucson.

In 1922 mining men Fred Sutter and W. B. Gohring drew Lemuel's attention to a group of twenty-two claims known as the 79 Mine, near Hayden Junction, from which they had extracted nearly a half million dollars worth of lead. Located on a limestone formation cut by porphyry dikes, the property had merit. A shallow shaft was sunk into an oxidized and silicified belt carrying a little copper, in which layers of lead carbonate were encountered varying in thickness from a few inches to four feet. The ore had been worked laterally via drifts for about 250 feet, before being cut off by block faulting. Believing the ore zones could be easily picked up again, Lemuel informed Chisholm and Bardon of the prospect of adding the 79 Mine to their mining properties.

The depression of 1922–23 and development of the Denn Mine dampened interest in the 79 Mine, which in the meantime was purchased by Los Angeles mining engineers R. C. Jamison and D. C. Peacock. They invited Shattuck to examine the property in November 1927. He again went over the claims, nosed about the drifts, and three months later directed the Shattuck-Denn geologist, L. T. McElvenny, to outline a core drilling program for Jamison and Peacock.[37]

Shattuck's inspection of the 79 Mine resulted in securement of $30,000 from the Shattuck-Denn Corporation to carry out exploration under direction of McElvenny, who opened an orebody 177 feet long by 42 feet wide, assaying 26 percent lead and 4 percent copper. By fall 1928 the mine was making money enough to enlarge and modernize the surface buildings and straighten the shaft.[38] By end of November Peacock and Jamison's attorneys were preparing to incorporate the company under the laws of Delaware. An issuance of treasury stock, taken mostly by directors of the Shattuck-Denn, raised $100,000, and set the stage for optioning of the 79 Mine by the Shattuck-Denn Corporation.[39]

While considerable exploration was carried out at the 79 Mine over the next several years, twelve years would elapse before the mine was put under the Shattuck-Denn corporate umbrella. The delay was caused by yet another financial debacle.

In both a figurative and literal sense, spirits ran high in Bisbee at close of 1928. As it had for the past fourteen years, the mining camp ushered in the new year with liberal doses of bootleg and looked forward to a prosperous future — in January 1929 the price of copper stood at eighteen cents a pound and the world screamed for the red metal. Every mine in the district was producing. Miners' wages advanced three times during the first quarter of the year, for a whopping raise of a dollar per shift. Restaurants were filled around the clock, and stores chalked up record sales. Vacant houses were a thing of the past, and roominghouses were full.

With money to spend, Bisbeeites were unconcerned when they read of the October 24 plunge of the stock market, the result of large sales of Kennecott Copper and General Motors stocks. When the market rebounded the next day, they passed it off as merely year-end profit taking. What people in the mining camp didn't know was that New York bankers had stabilized the market with an infusion of $240 million, which did little to dispel investors' doubts and fears. On October 28 the market dropped 12.82 percent and continued to fall, driving down equities $40, $50, and even $60 per share.

In a letter to his father, Mark Shattuck dismissed the plunge. "We are going to have a good market this winter nevertheless," he wrote.[40] His son's comment must have caused Lemuel to grimace, for he had been in New York attending a Shattuck-Denn board meeting and witnessed the confusion on Wall Street between October 24 and 28. "Most of the people there," Shattuck wrote, "acted like lunatics and a good many of them went to Trinity Church at the head of Wall Street to pray. I expect the longs prayed for the stock to go up and the shorts prayed for them to go down."[41]

Although business in Bisbee was unaffected by decline of the stock market, despite a drop in price of copper from 23 to 18 cents, Shattuck knew it was only a matter of time before the debacle's shock waves struck the town. The Miners and Merchants Bank had considerable money loaned to people investing in stock, and he stuck close to town, making sure his customers maintained proper margins or liquidated their stock if they did not. He could not afford to lose as much money as he had during the panic of 1907.

As a banker and president of one of the West's largest mining companies, Shattuck was on the inside track of the financial world. He was not optimistic. "I believe the market will go still lower," he wrote to Wiley Fitzgerald, manager of the Shattuck farm at Somerton.[42] While business conditions in Bisbee were actually "better than normal," Lemuel knew that "it all depended on the price of copper. If copper stays around eighteen cents we will be all right."[43]

An aging Lemuel Shattuck, pipe in hand, sits on running board of his 1934 Ford coupe about 1936. To his right is John Meadows. At Lemuel's left is John's sister, Belle, and her niece from San Antonio, Texas. William Lutley sits on front fender. (Fathauer collection photo)

TWELVE

THE GREAT DEPRESSION

AS Lemuel feared, the stock market dropped further in January 1930, heralding the onset of a worldwide industrial malaise. Commodity values evaporated. Credit dried up in America and Europe. Business slowed to a snail's pace. Copper, which hovered at eighteen cents a pound for months, dropped to fourteen cents a pound in April 1930, then slumped gradually during the succeeding six months, until it reached 9.59 cents per pound in October 1930. By end of December 1930 four million people were out of work in the United States.

The Depression caught metal producers with huge stockpiles of copper, which could not be moved to either domestic or foreign markets. American producers valiantly tried to bring production into line with consumption. They succeeded in restoring copper to twelve cents a pound early in 1931. The price could not be maintained. High grade copper deposits in Africa, South America, and Canada, worked by cheap labor, producing metal at less than five cents a pound, fed the glut of metal, forcing price of copper into a free-fall. The law of supply and demand had at last caught up with American mining companies.

To offset declining profits, companies reduced wages and shortened work schedules. A wage cut of 9.1% was established for all Arizona copper mines in October 1931. Production was halved. The Calumet and Arizona Mining Company and Phelps Dodge worked their Bisbee mines fifteen days a month, and ran the Douglas smelters at fifty percent capacity. The Shattuck-Denn shut down altogether.

Regardless of stringent cutbacks, some mining companies folded. Others merged. The Calumet and Arizona Company had enormous ore reserves and little cash. An opposite situation existed with Phelps Dodge; its ore reserves at Bisbee were depleted, but the company treasury was full. Late in September 1931 the two companies merged, leaving two mining companies remaining in the Warren District: Phelps Dodge, and the Shattuck-Denn.

The merger insured the survival of Phelps Dodge, but it hurt the workingman of Cochise County. Consolidation eliminated jobs. Supervisory help, as well as miners were laid off. There had once been two hospitals in the district, now there was one. Phelps Dodge shut down its Douglas smelter in favor of the more modern C&A facility. The change over eliminated 150 jobs at the smelter alone.

By end of 1931 three-quarters of Bisbee's work force was unemployed. "So many men were discharged from the mines, that half the houses in Bisbee are vacant," Lemuel observed.[1] The affect was devastating to merchants. Stores and restaurants closed. Single miners drifted off in search of jobs. Married men with families, who owned homes, lounged around town, destitute. Men who rented quarters were unable to pay their rent. Boardinghouse owners were powerless to collect, and it was useless to evict tenants. Buildings that had previously housed prosperous businesses, stood vacant, windows boarded up. The Shattuck-Schmid building, which sheltered the Brown Clothing Store on its ground floor, and offices of professional men upstairs, was empty.[2]

It was the worse of times for the mineral industry. Every mining district was shut down. Most hardrock miners had only one skill: wresting ore from the earth. With nothing else to do, they idled away the hours, the weeks, and months, subsisting as best they could. The destitute were fed by charitable organizations in Bisbee that somehow managed to collect about $5,000 monthly from employed individuals and the few surviving businesses.[3]

The two remaining companies contributed to the relief effort:

Phelps Dodge provided $1,400 monthly, the Shattuck Denn $250.[4] Together, they maintained sixty families of loyal employees who had worked in the mines for years. During 1931 the Red Cross shouldered some of the burden by aiding 1,700 persons, including 1,500 children. The Cochise County Unemployment Committee, formed later that year, eased the situation.[5] In 1933 these, and other agencies in Cochise County, raised over a quarter million dollars[6] to maintain soup kitchens for transient single men, and distributed food and clothing to needy families.[7]

Nevertheless, there was starvation in Cochise County. In March 1932 sixty-nine children were treated for malnutrition in Bisbee. Unemployed miners slaughtered a prize Hereford bull belonging to Frank Brophy. "I killed the bull because my wife, children, and I were hungry," confessed John Fonty.[8] "Conditions in Bisbee have never been as bad as at this time," lamented Shattuck.[9]

For a generation two-thirds of Arizona's industrial income came from copper. Because the state's forty-nine metal producing districts were the most profitable market for farm and range products,[10] cessation of mining forced farm and beef prices to an all-time low, bankrupting Arizona's cattlemen and agriculturalists. "All the ranches around here are broke," wrote Shattuck, "There are lots of ranches for sale in this vicinity, but they are no good, and no one has been able to make a living on them farming."[11] What manufacturing there was in the state also disappeared.

As both banker and mining company executive, Lemuel knew the statistics: a fourth of Arizona's population, exclusive of Indians, directly or indirectly depended upon the mining industry for a livelihood. In 1930 mining property constituted 39.78% of total assessed value of property within the state, and mines contributed forty percent of state revenue. To Shattuck, and every other resident of the Warren District, restoration and preservation of the mining industry was of paramount importance. A grassroots movement was forming in Cochise County to counteract what was considered the cause of the slump in mining — cheap foreign copper. Because of his knowledge of the copper industry, political and financial ties, the movement jelled around Lemuel Shattuck, and

Hoval Smith, former chief engineer of the Junction Development Company and an outspoken Republican who advocated a tariff on foreign copper as early as 1926.[12] John G. Flynn, Shattuck Denn general manager and president of Bisbee's Chamber of Commerce, made it a trio. The three would take the battle to the political arena.

Their first objective was to enlist the support of Governor George W. P. Hunt, then in his sixth term. That would be no easy task, for the governor was a "free trader," not always friendly to mining interests. As unemployment was rising and Arizona's economy was lagging, Shattuck and his colleagues sought to gain Hunt's support indirectly by organizing a conference on the "copper depression."

On May 11, 1931, men from every mining district in Arizona converged on Phoenix to discuss in a forum at the Hotel Adams, the plight of the state's mining industry. The press, alerted to the event, was there to channel a mountain of statistics to the reading — and voting — public. In a week-long series of meetings, mining men painted a dismal picture. In the past, when ore bodies were virgin, high grade, and near the surface, the copper industry had experienced periods of depression, which had been weathered through good management and improved extraction and refining methods. Each time, the industry emerged more efficient, able to produce copper at a cheaper price. By 1930 the situation had changed. Deep mining and a lower grade of ore had offset the benefits derived from improved techniques.

In the past American mining companies supplied the world with copper. Now, they were confronted with competition from Canada, South America, and Africa — areas which had the advantages of enormous rich deposits, cheap labor, low taxation, and the accumulated knowledge of American mining techniques. The African Roan Antelope Development, for example, could deliver copper to both European and American ports at under five cents a pound. The conference hammered home the fact that such mines were producing a copper tonnage far in excess of world demand, and that excess was destined for the United States and the American consumer. The conference gloomily predicted that the domestic copper industry, producing metal at a cost of from eight to ten cents a pound, would be forced out of business unless a tariff was enacted equal to the difference between the cost of domestic and foreign production.[13]

The conference and its attendant publicity focused Governor Hunt's attention upon the issue. He may have been an advocate of free trade, but he was a politician who recognized what appealed to the public. On May 20, 1931, he appointed an Arizona Copper Tariff Commission composed of Cleve W. Van Dyke, head of Van Dyke Copper, a company with properties in the Miami District, Sam Morris, an attorney in Globe, Bisbee banker Lemuel Shattuck, John G. Flynn, manager of the Shattuck Denn Company, and J. W. Strode, secretary to the governor. Van Dyke chaired the group, Strode acted as secretary.[14]

In the fashion of a true politician, Hunt picked up the ball by declaring May 25 to 30 "Copper Tariff Week." At a banquet on May 28 he introduced his Copper Tariff Commission to the general public, and announced the aims of the group: "to organize the move for a tariff on imports of the red metal into the United States."[15]

The commission embarked upon the task of saving Arizona's copper mining industry from extinction without any funding whatsoever. Their reward, Hunt promised, would "lie in the satisfaction that must come from the knowledge of having performed a valuable and useful work for Arizona."[16] Compilation of data was the first course of action, followed by discimination of information. Financed by the Miners and Merchants Bank, the Shattuck Denn, Van Dyke, and Miami Copper companies, Cleve Van Dyke and Hoval Smith hit the speaking trail, armed with facts gathered from the mining industry, world trade organizations, and Western industrial congresses.[17] The two men addressed civic clubs and chambers of commerces throughout the Western states. At the same time, the Commission pressured Senators Carl Hayden and Henry Ashurst, and Congressman Lewis Douglas to introduced a tariff bill, or to tack it onto an existing bill being considered by Congress.

Ashurst and Hayden endorsed such a bill. Lewis Douglas at first had reservations. He was philosophically opposed to tariffs, which seems odd in light of the fact that he was the son of James "Rawhide" Douglas who developed the United Verde Mine at Jerome, and a grandson of the man who brought the Warren District into production. A mine owner himself, Lewis believed that American copper producers had encouraged foreign production by demanding artifically high prices. Collapse of his state's economy, and pressure from constituents and the Arizona Copper Tariff Commission, caused him to revise his thinking.[18] In January 1932 Douglas introduced a special joint resolution to the Seventy-second Congress calling for a duty of 4½ cents per pound on copper imports. Opposition from free trade advocates quashed the resolution. Loss of one battle does not always determine the outcome of a war. A little "horse trading," as Shattuck put it, might yet gain passage of the tariff.

Working in the backgound, Lemuel contacted Southern California oil men, seeking to tie the tariff on metal to a tax on foreign petroleum then before the United States Senate Ways and Means Committee. To drum support, Shattuck put Cleve Van Dyke, "without a clean shirt or even a handbag," on a train for Washington, D.C., with instructions to make it known to the oil lobby that there was a "public demand here [in Arizona] that our delegation in Congress ally themselves with the oil people, or anyone else. They are rabid down here and are not only willing but practically demanding that our delegation vote against everything until we get a tariff. They are in misery and they either want to be helped out of it or have lots of company in their misery."[19]

In the meantime, Hoval Smith went to the lair of the Senate Finance Committee. In April 1932, he spoke to the 102nd Annual Conference of the Mormon Church, convened in the tabernacle at Salt Lake City. Smith's speech brought not only a standing ovation, a blessing by the President of the Church, it converted to the cause of copper Utah's Senator Reed Smoot, chairman of the powerful Senate Finance Committee.[20]

Copper tariff advocates still had to convince the House Ways and Means Committee. Eloquent pleading by Lewis Douglas and testimony from representatives of other copper-producing states eventually did the trick, and on June 6, 1932, both houses agreed to include in the revenue bill a duty on copper of four cents per pound, effective for two years.

Newspapers throughout the West proclaimed Lewis Douglas champion of the mining industry. The congressman was "entitled to a full measure of credit for the successful termination of the copper tariff campaign," acknowledged Cleve Van Dyke.[21] In actuality it was Shattuck money, and the silver-tongued oratory of Hoval Smith that won the day. The campaign, however, was not over.

The tariff, however, had no effect on the price of copper. There was too much red metal on the world market, which gave proponents of the tariff ammunition for seeking a permament tariff at the expiration of the law. Some even wanted to boost the tax to as high as ten cents a pound. For those reasons the Commission was kept alive,[22] and W. W. Lynch, United Verde sales manager, posted to Washington, D.C., to keep the issue before lawmakers. Meanwhile, Shattuck, Smith, Van Dyke, and Flynn organized benefits to raise funds for the Commission, and to publicize the plight of the copper industry.

Assumption of the presidency by Franklin Delano Roosevelt on March 4, 1933, changed forever the economic course of the United States, and the non-ferrous metal industries. Taking advantage of a clause in the Industrial Recovery Act, authorizing him to enter into agreement with employers, the President on July 27, 1933, launched an emergency drive against what he termed the "forces of depression." He established the National Industrial Recovery Administration (NIRA), conceived to be a "new expression of partnership between industry and government," expressed as a voluntary agreement between employers and the President to set maximum hours and minimum wages of employees for the remainder of the year, pending approval of codes of fair competition.[23] Basic codes for all American industries were to be gathered, without delay, by the NIRA.

The copper industry wrangled over the code for nine months; primary producers wanting one thing, custom smelters seeking another. Both groups submitted their own codes. The NIRA, however, wanted a single code by which the entire industry would abide. An imperfect code was finally promulgated and submitted on November 14, 1933. Labeled a "Code of Fair Competition for the Copper Industry," it set only a duty on fabricated copper, and established production quotas for mining companies. Each primary producer, about ten companies, having a productive capacity of not less than 15,000 tons per annum would be allowed to produce at a rate of twenty percent of capacity, plus 2,500 tons per annum. Submitted on November 14, 1933, the code brought protests from smaller mining companies in the West with less productive capacity. Lemuel Shattuck felt the code favored the ten large companies and closed the door to small producers. Labeling the clause specifying disposal of copper inventory "indefinite and uncertain,"[24] Shattuck vehemently opposed the code. "It will not benefit the communities dependent upon copper mining or the stockholders of independent copper mining companies. In fact, it . . . will increase the distress in the mining communities, and may ruin the interests of both our employees and stockholders."[25]

Shattuck, Van Dyke, Smith, Flynn, and others in Arizona turned to their congressional delegates: Hayden, Ashurst, and the state's first congresswoman, newly-elected Isabella Greenway, wife of deceased copper magnate John Greenway. At hearings on March 12, 13, and 20, Arizona's three representatives voiced opposition to the code, stating that it favored large mining companies, and did not consider the welfare of employees. Cleve Van Dyke suggested that copper production be allocated by state, which would "materially increase employment by reviving the industry." Ashurst and Hoval Smith urged the President to promulgate a code if the industry could not agree upon one. Congresswomen Greenway requested a code that would "really put copper miners back to work."[26] On April 21 the copper industry, wishing to avoid having a code forced on it, reluctantly accepted essentially the guidelines proposed earlier by the NIRA.

Except in case of national emergency, the code did not fix prices. It froze surplus stock, and set production quotas for all companies on newly produced metal called "Blue Eagle" copper. And it fixed hourly wages for all employees, underground and on the surface. In the Southwest underground workers received forty-five cents per hour, surface men thirty cents per hour.

The copper industry had experienced the worst period in its history. Metal prices had gone from twenty-four cents in 1929 to four cents in 1933, bankrupting mining companies, and threatening to turn once prosperous communities into ghost towns. The battle to save the industry, initiated in Bisbee, but fought in Phoenix, New York, and Washington, D.C., had been long and bitter. In contrast to past debacles, when the copper industry had emerged much stronger, a weak, badly shaken, industry had weathered this depression. "Convalescent" was the term used to describe its condition.

In desperation to save their companies and communities, Shattuck and other owners of independent mining companies initiated the fight for a tariff against imported copper. That battle resulted in a deficient code, which Lemuel and his colleagues hoped to remedy in July 1934. Election of Roosevelt shattered hopes of increased tariffs as a way of putting mining back on its feet. Instead of establishing tariffs, the New Deal demanded fair practice codes, and then dictatorially ordered American industry back to work.

As a staunch Democrat Shattuck supported Roosevelt, but objected to New Deal methods. Lemuel thought the NIRA had strictured the copper industry by arbitrarily setting quotas, fixing prices and wages. It had done the same thing to banking. For that story we will have to backtrack to 1929.

Unemployment and the erosion of Arizona's industrial base wreaked havoc with banks. Deposits declined by half. Commercial loans fell by two-thirds, and the loan-deposit ratio dropped by forty-one percent. There were forty-six banks in Arizona in 1929. Over the next three years, that number decreased by over fifty percent.

Signs of an impending economic crash were apparent to Shattuck long before Black Thursday. Five years of good times, as every banker knew, was pushing economic luck. Speculators were playing the stock market like roulette, buying and selling equities on margins impossible to cover in event a downturn forced a call. By end of 1929 circumstances were in place for a major debacle, and astute bankers, including Shattuck, had prepared for it.

Early in 1929 Shattuck and other directors of the Miners and Merchants Bank, restructured their institution's policy to provide maximum liquidity. Loans for purchase of stock would be made only to members of Bisbee's "gambling" fraternity, namely brokers Overlock and Stevens. With exception of a few real estate loans in Nogales, Tucson, and Phoenix that would net eight percent, no real estate loans were made. Collateral for commercial loans was increased from two to three dollars value for every dollar loaned. The Bank of Bisbee followed a similar policy. Both banks, each with nearly four million dollars in assets, were in a strong position in 1930 when the depression struck.

Bisbee's bankers had only the depressions of 1907–10 and 1922–23 to judge the severity of this downturn. They never expected copper — the mainstay of their town's economy — to lose two-thirds of its value. When production was curtailed, and wages slashed, bank deposits correspondingly dropped. Earnings of the Miners and Merchants Bank for 1930 were $83,000 less than the previous year.[27] General unemployment brought defaults. The Miners and Merchants Bank wrote off $107,000 in uncollectable loans in 1931.[28] The depression was unmatched in severity, but Shattuck was tenacious, determined to weather the storm. "There seems to be a great depression in business all over the United States; in fact, all over the world at the present time. But we have experienced the same condition before, and if we are careful and keep working, we can tide things over."[29]

Many bankers were unable to "tide things over." Collapse of farms prices forced the closure of three Yuma Valley banks, and impacted directly on Lemuel Shattuck. The Caruthers State Bank, at Somerton, closed February 28, 1930, tying up the account of Wiley Fitzgerald, manager of the Shattuck farm.[30] A year later, Security Trust and Savings Bank, Yuma's oldest institutions, went into receivership. As Emil C. Eger, director of the institution, handled collection of rents and notes for Shattuck property in the valley, the bank's demise hit Lemuel's pocketbook. Three months later E. F. Sanguinetti's Yuma Valley Bank went under.

As banks faltered, their presidents and directors looked to those shrewd individuals who had resources enough to weather the depression. Heads of several Tucson banks came to Bisbee to plead with Lemuel for help in saving their institutions, but there was a limit to what he could do. Tucsonan John Murphey and his wife had homesteaded hundreds of acres in the Catalina Foothills. He needed cash, and offered to sell Shattuck some of the property at ten dollars an acre. Lemuel brought the proposition before directors of the Miners and Merchants Bank. They turned the offer down, saying "it was land that wouldn't support anything but jack rabbits." Ironically, by the late 1970s this land was selling for tens of thousands of dollars an acre.

Lemuel was confronted with opportunities at every hand. In December 1930 aging cattle baron and banker, B. A. Packard, offered to sell Shattuck controlling interest (720 shares of stock) in the First National Bank of Douglas, as well as $75,000 worth of stock in the Valley Bank and Trust Company at Phoenix.[31] Lemuel thought long and hard about Packard's offer. Ninety days later he wrote the old man, "I do not care to buy any more banks, as I have all I can attend to with our little bank in Bisbee."[32]

Failure of the Yuma Valley Bank during summer of 1930 presented Lemuel another opportunity. On July 4 Sanguinetti sought Lemuel's aid in reopening the bank. "The closing of the Yuma Valley Bank was a serious blow to our community, and we are now very much concerned in securing a good substantial financial institution. . . . Many of the depositors . . . stated they would be willing to make a sacrifice of from 15% to 20% of their deposits as a bonus, if the Yuma Valley Bank could be reopened."[33] Memories of Winterhaven still burned in Shattuck, and there was no sorrow in Lemuel's answer to Yuma's Merchant Prince. "Personally, I do not care to become interested in the bank, as I have all I can do to attend to our bank."[34] One man's misfortune can be another's op-

another's opportunity. Lack of banking facilities in the Yuma Valley gave Shattuck the chance to reestablish his financial presence in the Colorado River Valley.

Failure of the Yuma banks presented an irresistible opportunity for any bank with resources. In the depths of the depression few institutions were solvent enough to venture into a region totally dependent upon agriculture. The Citizens' National Trust and Savings of Los Angeles long financed Southern California agriculture, and sought to expand into Arizona. In spring 1932 it applied to the Comptroller of Currency for a national bank charter for a Yuma branch. Shattuck's Miners and Merchants Bank, which had financial ties to the region dating back to the days of the Winterhaven Commercial Company, beat the Los Angeles bank to Yuma by several months.

Lemuel had acquired a permit extending his bank's charter to Yuma, and purchased for $15,000 the building formerly occupied by the Security Trust and Savings Bank. Late in May he sent cashier I. F. Burgess, bookkeeper Franklin Oliver, and teller R. T. Edwards to Yuma to prepare for the opening of the branch. Bank employees would alternate between Yuma and Bisbee until the bank was permanently staffed. The branch was formally opened on June 10. In that age of tight money and worthless paper, the Miners and Merchants Bank exercised the utmost caution. Spencer Shattuck and I. F. Burgess spents days pouring over books of the Yuma County treasurer's office to determine if the county was solvent before cashing county warrants.[35]

Opening the Yuma branch gave the impression that the Miners and Merchants Bank had been unscathed by the depression. Nothing could be furtherest from the truth. In actuality Lemuel Shattuck was fighting a battle that had not only local and state ramifications, but impacted on national economic issues. On a personal level he was concerned about preserving his bank, and maintaining the value of his mining company. On a local level, he fought to prevent Bisbee from ending up a virtual ghost town as Tombstone had. And he was aware that Arizona was inextricably linked to copper in more ways than one.

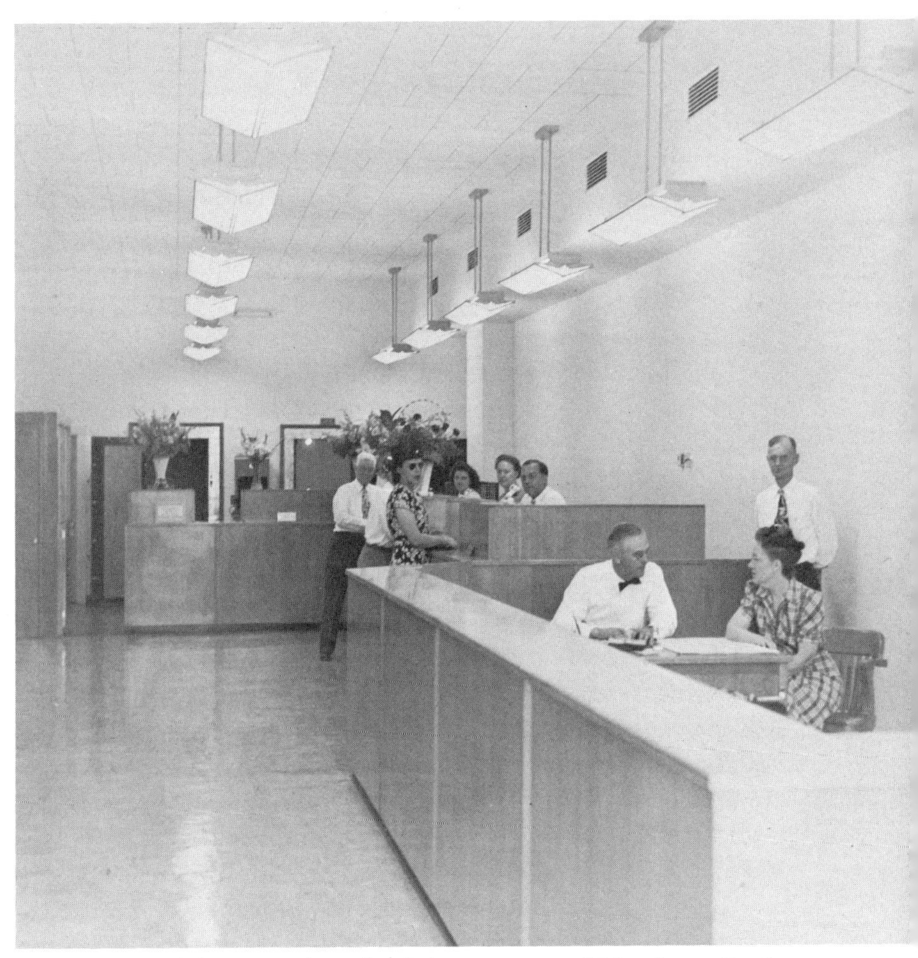

The interior of the Yuma branch of the Miners and Merchants Bank.
Spencer Shattuck talks to woman patron.　　　　(Fathauer collection)

The ninety-five percent decline in copper over a three year period affected all sectors of Arizona's economy. Farm prices dropped, manufacturing ceased, services suffered. General unemployment syphoned savings and created loan defaults, cutting to the heart of banking. One-by-one banks closed, and those remaining open were near a state of insolvency.

Insolvency was not just an Arizona problem, however. The nation's entire banking system was on the verge of collapse. Between 1930 and 1933 more than 6,000 banks with $3.5 billion in deposits disappeared from the scene. Bank holidays, called by state governments to prevent runs on the institutions, became common. In October 1932 Nevada declared a twelve-day moratorium and then extended it by three weeks. On February 4, 1933, Louisiana closed its banks, and ten days later Michigan called a ten-day bank holiday, followed a week later by Maryland. By end of the month, seven other states suspended banking. Governor James Rolph of California declared a moratorium on March 1, 1933.

California and Arizona was linked by a transaction network. Arizona banks had considerable portion of their funds and credits in California institutions. Interruption of the freedom of banks to draw upon their reserves spelled calamity. Runs on Arizona banks would surely result when news of the California bank moratorium leaked out. To forstall that possiblity, Arizona's Governor Benjamin B. Moeur, wielding emergency executive powers, rushed a moratorium bill through the legislature in fifty-five minutes. Telegrams were dispatched to every banker, ordering all financial institutions closed for three days. The next day the governor made the order mandatory, and the state superintendent of banking followed with a statement that Arizona's banks were "in good shape."

Most bankers complied with the governor's order, and closed their institutions. Bisbee's men of finance were a stubborn lot. James Douglas, who controlled both the Bank of Bisbee and the Bank of Douglas, believed the law passed by the legislature did not give the governor mandatory powers to close the banks, and concluded to keep his institutions open. This put the Miners and Merchants Bank in a bind.

The institution was the largest bank in Cochise County and fourth ranked in the state, and Lemuel Shattuck had worked hard to put the Bank in shape to weather the depression. He had cut down loans, called in notes, and built cash reserves. And he had courageously moved into the Yuma Valley. The governor's telegram hit like a bomb. To close the bank, Shattuck felt, would benefit the Bank of Bisbee and do a disservice to the struggling Warren District. On March 3, Shattuck acknowledged Governor Moeur's telegram as:

. . . sufficient authority for our bank to close. However, the Bank of Bisbee and the Bank of Douglas people . . . state that they are advised by their attorneys that in their opinion the law passed yesterday did not give you mandatory power to close the banks and they have concluded to keep open. Such being the case, for obvious reasons, we cannot very well close while they remain open.

I might add that I would very much like to keep our Yuma branch open owing to the sentiment of the people down there in regards to banks, five of them having closed there in recent years, and we are doing everything in our power to regain their confidence. I feel that by closing our Yuma branch, even for a few days, it would cause us to lose what confidence we have gained.[36]

Until he received clarification of the governor's actions, Shattuck kept the Miners and Merchants Bank and its Yuma affiliate open. The bank holiday nevertheless increased anxiety in Cochise County, already jittery from the decline in mining. People pondered whether or not to withdraw their precious savings. Runs on the banks were a distinct possibility. The next day Lemuel Shattuck and M. J. Cunningham moved to quiet fears, and restore public confidence, by publishing a joint statement "that banks in Bisbee have been preparing for present stringent conditions for many months, and that they now have—and have always had—sufficient cash money to care for any demands which may be made upon them." The two bankers assured Bisbeeites in particular and Arizonans generally "that the two . . . banks have been doing business . . . more than thirty years, that they have gone through several depressions, and that they expect to continue serving the Bisbee district in the usual way for many years to come."[37]

Angered by the rebellious Bisbee banks, Governor Moeur ordered his proclamation posted over their doors. The posters carried a postscript suggesting that their removal was sufficient cause for the National Guard to be called out to enforce the order. The bankers were not intimidated. Both Shattuck and Cunningham believed their community was "more or less isolated" and saw no reason for "discommoding patrons of the banks." The next day, however, Lemuel reversed his stance and closed the Miners and Merchants Bank, both at Bisbee and Yuma. Why he decided to comply with the governor's order is not clear.

His change of heart may have been due to the fact that he had always worked for the good of Arizona and the Warren District. To keep the bank open would increase the governor's difficulty in stabilizing the state's banking system. Lemuel may have feared repercussions from state banking authorities. Then too, Shattuck may have seized the chance to force out the long standing competition. There is no hint in Shattuck's correspondence why he reversed his stand.

Shattuck's actions insured a run on the Bank of Bisbee. Rather than risk that catastrophe, Douglas ordered his bank closed on March 5. The next day, President Franklin Delano Roosevelt declared a national banking moratorium. On March 9 the Emergency Bank Act was passed, confirming the President's proclamation, and the following day the moratorium was extended for an indefinite period.

Roosevelt's Emergency Banking Act went far beyond mere suspension of bank transactions. It placed an embargo on gold exports, authorized clearing house certificates as a medium of exchange, appointed conservators for insolvent national banks, and liberalized powers of the Federal Reserve to issue circulating notes. On March 12 Roosevelt delivered a masterful "fireside chat," explaining reasons behind the moratorium, and what was being done to correct the national banking crisis. Every bank in the country had to be inspected by federal examiners before they were licensed to reopen. One-by-one banks were opened between March 12 and 15. The Miners and Merchants Bank and the Bank of Bisbee received their authorization to open on March 15, 1933.

While only four Arizona banks were declared insolvent, the moratorium was a close call for some. The holiday caught the Riggs Bank in Willcox with too little cash and too much "paper" on small farms and ranches in southeastern Arizona. Federal examiners prohibited the bank from opening. Although the Riggs brothers had disassociated themselves from the bank years before, they sought to bail out the institution because it "bore the family name." Turning to Lemuel Shattuck, they offered their ranches as collateral for a loan of $75,000.

Shattuck could not turn his back on this situation for several reasons. Failure of the Riggs Bank would wipe out many small farmers and ranchers, who also owed money to the Miners and Merchants Bank. Then too, the Riggs and Shattuck had long been friends. For years the Miners and Merchants Bank had maintained a $50,000 line of credit for running of the Riggs ranches. Never defaulting, the Riggs' credit was sterling. Lemuel drafted a note with the ranches as collateral, dug $125,000 from his personal savings, and sent it to Willcox via his attorney Jim Gentry. The Riggs Bank opened on time under the name of the Bank of Willcox. True to form, the brothers paid Shattuck back in two years.[38]

The moratorium halted hoarding, stabilized America's banking system, and signaled the beginning of the long climb out of the depression. Arizona bankers, however, faced another crisis of major proportion—the intangible property tax. To pick up lagging state revenue, Governor Benjamin Moeur resurrected and revised a forgotten 1913 law which taxed banks on capital stock rather than on real property or other assets. Renamed the Intangible Property Tax Act, the law was modified to levy taxes on the "fair value" of bank stock, "to be determined by an analysis of reserve and surplus accounts, and undivided profits, including dividends and market price."[39] The law required payment of $15 per thousand dollars assessed value to stockholders, and forced banks to pay taxes under a state rate on all real estate held by banks, in addition to paying city and county real estate taxes. The law also made banks tax collectors for monies on deposit. In the case of the Miners and Merchants Bank, that amounted to $6,000 per annum. The intangible property tax created an uproar from the banking community.

No sooner had the governor submitted the modified version of the law than Lemuel Shattuck protested its constitutionality. On June 7, 1933, he complained to state senators Dan Angius and John Riggs that "If this law should be enacted we believe that 90% of our depositors would withdraw their money from all the banks in the State and keep their profitable accounts either in the Post Office or in banks in other states, and would practically put all banks in this state out of business." In addition, Shattuck pointed out, "this law also puts an additional tax on the mining companies who are now bearing the greater part of the tax burden. It would make all Arizona stockholders pay a tax to try and hold their stocks to get even."

"You boys," Lemuel urged, "should kill this bill in the Senate by all means, and . . . endeavor to get a fair and equitable sales tax which would apply to everybody, farmers as well as anybody else."[40]

The modified law was enacted that summer despite opposition and lobbying efforts from mining and banking interests, causing an uproar from financiers. "No honest man can remain in the banking business," declared James Douglas. Already irritated by New Deal policies, the intangible property tax law was the last straw for that irascible banker. In mid-November 1933 he voluntarily closed his Bank of Clemenceau in Jerome. Convinced that "the banking business in . . . Arizona is handicapped by being compelled to carry a larger burden of taxation in proportion to its wealth than other classes of property," Douglas prepared to liquidate his Cochise County banks. When he announced his determination to get "out from under" the Bank of Bisbee, Frank Brophy offered to buy the faltering but viable institution.[41] Fearing such a sale would open the door for Shattuck to acquire the institution, Douglas announced on February 1, 1934, that the Bank of Bisbee would pay off its depositors and voluntarily close on its 34th anniversary, February 19.[42]

Lemuel Shattuck did not like the restrictions and intervention of the New Deal any better than Douglas, and the intangible property tax was repugnate. But closing solvent banks in protest was an act damaging to community interests. A different man than mercurial James Douglas, Lemuel had promoted and protected the Miners and Merchants Bank for thirty-two years. He would fight the Intangible Property Tax Act.

Lemuel paid assessments against the Miners and Merchants Bank, and then directed Spencer to file suit questioning the validity of the law. Litigation dragged on for a year, finally reaching the state supreme court. On November 27, 1934, the court declared the law unconstitutional—a major victory for not only the Miners and Merchants Bank, but for the state's financial and mining community as well. Although the Bank of Bisbee eventually reopened as a branch of the Bank of Douglas—Brophy having saved the institution through a complicated manipulation of stock—the Miners and Merchants Bank was the sole bank in Bisbee for seven months, a period of intense activity for Lemuel.

With ending of the moratorium, depositors returned in droves to deposit cash and gold, hoarded for months. Resumption of mining under NIRA guidelines pumped cash into the community, as did other New Deal projects. The Civil Works Administration approved $352,300 worth of construction projects for Bisbee, which put back to work two out of every five persons on the community's relief rolls. Their wages were not high, only fifty cents an hour, but when the checks were cashed it was adrenalin to the business community.[43] For the first time in years Bisbeeites demonstrated a certain exuberance, added to in no small degree by repeal of prohibition. In April of 1933, thirteen Bisbee establishments were given permits to dispense beer.[44] Repeal of prohibition fostered nostalgia for the days of the saloon. Reminding Anheuser Busch he had sold more than a million dollars worth of their keg beer before Arizona went dry, Lemuel Shattuck requested a distributorship for the Warren District. As Isabella would have done, Mary O. punctured his dream by telling Lemuel he was too old for the saloon trade. He agreed. At sixty-seven he did not have the vitality he once had.

Lemuel's fatigue stemmed from more than just business worries. Family problems ate at him. Always tightly knit, the Shattuck family was now dispersed. Dorothy and Isabel were attending school in Chicago, and Mark and Warner resided in California. Only Spencer had returned to Arizona to become his father's right-hand man. Sons on the West Coast caused Lemuel some concern.

Shipping the two eldest boys to Mexico to avoid World War I damaged the relationship between Lemuel and Warner, and Henry's disappearance destroyed it. Lemuel's remarriage further wounded Warner. Why he objected to his father's second marriage is not fully understood: it may have been caused by thoughts of reduced inheritance, or merely Warner's inability to accept a stepmother. Ill-will was there nevertheless. He never wished his father happiness as the other children did, nor visited Bisbee to meet Mary O. It was only through Mark, the family's little "red hen," that Lemuel was able to keep track of his eldest son.

Mark maintained a close relationship with all members of the family: looking after his alcoholic brother, fretting about his sisters, and corresponding weekly with his father. Lemuel never went to the West Coast without visiting Mark, and the two occasionally went deep-sea fishing, a source of great joy for both. Their visits, however, were marred by depression blues. In 1929 Mark invested $110,000 of his inheritance in Shattuck Denn stock. Drop of the market locked up the sum until such time as copper rebounded. Losses on oil stocks, and municipal bonds also eroded Mark's fortune to the point that he was forced to borrow money from his father. Mark's financial worries aggrevated a coronary condition originating from childhood rheumatic fever. Mark's letter of January 16, 1933, relating what he termed "a let down of the nerves," caused his father grave concern.

> I . . . was taking a shower bath when all at once everything turned black and I fell to the floor. I managed to drag myself to bed and I thought that it was going to be the end of me. . . The first days of my sickness were plenty bad. I threw up all day long and was so sick I wished I would die.

No case of nerves, this stroke put Mark in bed for ten days and temporarily paralyzed his left side. Although the impairment was transitory, he suffered spells of dizziness which altitude compounded. After 1933 he seldomed ventured to Bisbee.

Spencer Shattuck's college days provided Lemuel moments of uneasiness, which passed with his son's marriage and graduation. Coming out of Stanford with honors, Spencer stepped into the role envisioned for Henry — the apprenticeship for overseer of family enterprises. Much to his father's delight, Spencer threw himself into all facets of bank management. When the Bank of Bisbee closed, he spent long hours working as a teller, accommodating patrons as they shifted funds from one bank to another. And he aided his father in processing the mountain of loan applications from businessmen previously financed by the defunct bank, and from long-time customers whose businesses were on the mend. And it was Spencer who guided the Yuma branch of the Miners and Merchants Bank. By the mid-1930s Lemuel was taking full advantage of his young son's business ability and decisiveness.

Lemuel was sixty-four years old when the Great Depression struck, not a young man. Although he worked as always, a twelve hour day, and traveled extensively on business, by 1932 Shattuck was feeling his age. That spring he turned banking duties over to Spencer, and placed mine operations in J. G. Flynn's hands and went "bumming." As lazy as it sounds, the term meant a period of conferences with Shattuck-Denn board members in New York and Chicago, interspersed with brief periods of relaxation with his daughters, who were embarking on lives of their own.

Late in 1930 Isabel met Walter Fathauer, son of Theodore Fathauer, a Chicago lumberman and philanthropist who owned extensive timber properties in Arkansas. A graduate electrical engineer from MIT, Walter was an intensely serious person. He recognized Isabel Shattuck as "a charmer," who possessed "all the characteristics" that he liked in a person. What he called his "conservatism," however, prevented Walter from launching a whirlwind courtship. "After fifteen months of deliberation and introspection," he found himself "in the position of one who is madly in love." On January 11, 1932, Walter Fathauer requested Shattuck's permission to marry Isabel.[45]

Lemuel was aware of Walter's characteristics through correspondence with Isabel. The young man was steadfast, frugal, intelligent, witty, and knew where he was headed. Lemuel found Walter's directness refreshing, and felt his seriousness would counteract his daughter's "nervousness." Furthermore, being men of conservative views and German ancestry, Theodore Fathauer and Lemuel Shattuck instantly liked one another. Lemuel happily consented to the Fathauer-Shattuck union, and Isabel and Walter were married on April 2, 1932.

In the meantime, Dorothy had fallen in love with a law student and a friend of Walter's. It was music, not the glamor of a potential career in law, that attracted Dorothy to Ted Stensland. He had grown up in the world of music, his mother having sung opera in La Scala, Italy, and later, headed the Voice Department at the Chicago Music Conservatory. By January 1932 thoughts of marriage were in the wind.

Lemuel met Ted Stensland for the first time while attending Isabel's wedding in Chicago. He liked him, but suggested to Dorothy the marriage be postponed until Ted finished law school. Dorothy did not heed her father's advise and on April 23, 1932, two weeks after her father's departure, they married. A meeting of the Shattuck Denn board of directors in Bisbee prevented Lemuel from returning to Chicago for the wedding of his "baby girl." Instead, he wrote, "I will be thinking of you and hoping for your continued happiness." He followed with some fatherly advice:

I think that you understand that when a girl marries it is her job to help her husband, and as you are marrying a boy that is not yet established, you should be economical. . . .

If you will think and consider, you will probably realize that a wife can make or break a husband, and it is up to you to see that your husband is helped in such a fashion that he will amount to something in the future.[46]

Lemuel folded into the letter a draft for $500 "to spend wisely in starting house keeping."

The newly-wed couples set up housekeeping in small apartments on Chicago's north side. Money from their mother's estate, which had been put in trust by Lemuel, allowed Isabel and Dorothy to augment their husbands' meager incomes, and they lived comfortably in a world of economic chaos, attending college, entertaining, boating on Lake Michigan, and going to concerts and art shows. For the first time in their lives, Isabel and Dorothy felt close. In 1932 they finished their educations; Dorothy attaining a teaching certificate from the Chicago Music Conservatory, and Isabel a bachelor's degree in economics from Northwestern University.

Isabel's wedding offered Lemuel a joyous respite from the problems and tedium of running businesses in a state whose economy had collapsed. But the situation was improving. The glut of copper was dissipating, causing metal prices to climb toward nine cents a pound even before passage of the copper code. During summer of 1933, 900 Bisbee miners were put back on a full-time work schedule. Shafts were pumped and retimbered, stopes cleared and put in shape for production. The Junction shaft, which Phelps Dodge acquired through merger with the C&A Company, was deepened. The Douglas smelter, scheduled to be closed until September of that year, was reactivated in July. By January 1935 mines were once again producing metal according to NIRA quotas.

As the Junction Mine went, so did the Shattuck-Denn. Pumping of the Phelps Dodge shaft drained the Denn, permitting resumption of mining in October 1934. Production was not large, only 3,000 tons of ore per month, but the $14,000 per month derived from sale of the refined copper heralded the prospects of better days. With that in mind, Lemuel ordered the Denn shaft sunk to the 2,700 foot level to exploit several large ore bodies discovered by the doodle bug probes of Weldon Humphrey and his son prior to onset of the depression.

At nine cent a pound, the price of copper tottered at the break-even point, furnishing small compensation to the company. New Deal monetary policies brought a resurgence of mining through-out America, creating a flurry of trading on the stock market. When the Denn resumed production, its stock advanced from two to six dollars a share.[47] It had been a long battle just to reach this point, and Lemuel was content to see the reactivation of mining. There was a "livelier look" to the Warren District.[48] "I believe that conditions . . . are a little better than they have been for some time," Shattuck commented to an old friend in July 1934.[49] Sixty days later he and Mary O. left for the Hawaiian Islands, as he expressed it, "to get a long ways off for three or four weeks so as to forget about bum business."[50] The fatigue of battle clearly showed.

The Hawaiian vacation rejuvenated Lemuel and he went back to work with his usual vigor, paying weekly visits to the Denn Mine to inspect the progress of shaft sinking, which then was exceeding three feet per day. It was Lemuel's intention to sink the Denn to the same depth as the Junction Mine, the deepest in the Warren District.[51] Improved silver prices, thanks to New Deal monetary policies, caused Lemuel to once again think about financing mining ventures. In April 1935 the Miners and Merchants Bank advanced the 79 Mine $10,000 to retrieve silver ore contained in pillars left standing in the older portions of the workings.[52]

At the same time, Shattuck ordered exploration at a group of claims in Cave Canyon on the west side of the Huachuca Mountains, of which he held the major interest. Long known to contain lead, silver, and minor amounts of zinc, Lemuel hired Weldon Humphrey and his son to determine the extent of ore horizons on the property. Core sampling and magnetometer readings revealed sizeable lenses of ore laying below the water table in limestone strata capped by rhyolite. The tests pointed to a great expense in shaft sinking and pumping, perhaps $50,000 or more. Not wanting to spend such a sum, Shattuck turned to other mining companies in June of 1936, bringing in engineers from Eagle, Pilcher Lead Company, U.S. Mining and Smelting, and Kennecott. The hard rhyolite capping the ore-bearing limestone dampened interest. Everyone investigating the property felt that in light of current metal prices, the risks and costs of development were too high. Unable to move the property, Shattuck lost interest in Cave Canyon.

Lemuel had no better luck in Mexico. In 1918 Lemuel acquired, in partnership with Joseph Muheim, and Michael Cunningham, a gold property in the Huajicori District of Nayarit, called the Esperanza. Except for periodic development, revolution rendered the property impossible to work. Shattuck nevertheless paid the Mexican taxes to retain ownership for himself and his partners. By the 1930s political unrest had quieted down enough to allow development of the district, and attention was again focused on the Esperanza. In November 1934 Walter Frazier, owner of the neighboring Cori mine, expressed interest in the claims. Shattuck put a price of $35,000 on the Esperanza and opened negotiations through a Mexican lawyer, Abraham D. Ortiz.[53] Haggling went on for nearly two years and finally collapsed. In all probability, Shattuck would have closed the deal had it not been for waning health.

Late in September 1936 a heart attack struck Lemuel. Nelson Bledsoe, his doctor, ordered him to bed. For three weeks he remained inactive, irritated by his inactivity. "For the first time in my life I have been laid up on account of sickness," he wrote rancher Neil Erickson.[54]

Shattuck was out of bed in three weeks, but he had no stamina. When he complained to Bledsoe, the doctor was adamant: "you have to take it easy." Lemuel followed doctor's orders, working only a few hours in the morning, and napping and reading in the afternoon. Recovery was slow. "I . . . have been no good since I had that heart attack," he wrote Wiley Fitzgerald.[55] Unable to do more for his old friend, Bledsoe referred Lemuel to an El Paso heart specialist, who told Shattuck to "take it easy for a long time. You can not stay at the bank all day until your heart gets a whole lot better."[56]

Having lived life at a fast pace, Shattuck was impatient with his physical condition, and the limitations imposed by age, the latter accentuated by the death of several close friends. Baptiste Caretto, whom Shattuck met while working as a trammer at the Czar Mine in 1889, died of influenza on January 16, 1933, ending a friendship of forty-four years. Lemuel J. Overlock, who, with his brother Charles, ran the butcher shop next to the St. Louis Beer Hall, and later headed Bisbee's leading stock brokerage firm, died on June 13, 1934. Besides being a dear friend, Lemuel Overlock helped Shattuck engineer the financing of mining ventures. Shattuck sorrily missed both men; their passage brought home the inescapable fact that his life too was drawing to an end.

Actions between the time of his heart attack and spring of 1937 indicate Shattuck was concerned about his vulnerability. At Christmas of 1936 he forwarded to Isabel and Walter a savings account passbook in the amount of a thousand dollars, as a gift to his year-old granddaughter Mari.[57] He also presented large blocks of Shattuck-Denn stock to Dorothy, Isabel, Warner, and Spencer, and cancelled a note Mark owed for $5,000.[58] He was careful each of his children received gifts of identical value.

In hopes of speeding Lemuel's recovery, the El Paso doctor suggested treatment at Bad Nauheim, Germany, a center of European research in heart disease. Shattuck fatalistically accepted the advice: "I will have them go over me and take their treatments. Maybe they will help me and maybe they will do me no good. But anyhow, I will have tried."[59]

On April 18, 1937, Lemuel and Mary O. sailed from New York aboard the *S. S. Bremen*. He underwent five weeks of treatment at Bad Nauheim, consisting of a daily regime of pills and physical therapy in heated spas. Enough of his old vitality returned, that the Shattucks extented their European stay, "just bumming around looking at sights" in Austria, Czechoslovakia, Denmark,

Mary O. and a frail Lemuel Shattuck at Bad Nauheim, Germany, where he underwent a month of treatment for his heart condition.

(Fathauer collection)

Sweden, and Norway.[60] Lemuel returned to Bisbee feeling "very well."

Lemuel took up a half-day work schedule, going in the morning to his office at the Miners and Merchants Bank to pass on business matters. As ever, he delved into mine speculation. In spring of 1937 he considered investing in the Black Diamond property in the Dragoon Mountains, but turned it down because there was no depth to mineralization. That fall he engineered the sale of the bank's Yuma branch to First National Bank, a Transamerica Corporation affiliate.

The summer of 1938 Isabel and Walter asked him to accompany them on a trip up the St. Lawrence River to visit Dr. Wilfred Grenfell's health clinic and project in Newfoundland. He opted instead for a voyage up the Inland Passage to Alaska. It was a trip that would touch the soul of any mining man: pristine beauty and frontier conditions.

In early September, within days of his return from Alaska, while reading a newspaper on the sun porch of his Warren home, Lemuel became aware of pain and swelling in his right leg. Mary O. wanted to rush him to the hospital, but he would not budge until he read of the fate of the then-free city of Danzig, Germany, for Lemuel was concerned about the rise of Nazism. Admitted to the Copper Queen Hospital, he was diagnosed as having a blood clot. Medical practices of the 1930s were a far cry from those of today. There was no treatment to isolate the clot, nor medicines to dissolve it. As a result, Shattuck suffered a major stroke and lapsed into a coma. With every family member, except Warner, at his bedside, Lemuel died early in the evening of September 7, 1938.

At that moment Isabel recalled what her father had said years before. "It is easier to die than to live." In her father's case, she had to agree.

The beloved banker and miner was dead. His funeral two days later was the largest ever held in Bisbee. Bankers, politicians, mining executives, ranchers, and old-timers from all over the state attended the service in St. John's Episcopal Church. The entire town turned out. Those who could not find seating in the chapel stood silently in the street. It was a solemn procession that wended its way through Bisbee and Lowell toward the Evergreen Cemetery. Bisbeeites congregated along the way, removed their hats as the hearse passed by — their last tribute to a founding father of their town and a friend of the hardrock miner. Lemuel was laid to rest in the family plot, next to Isabella, within 200 yards of the Shattuck Denn Mine.

For days thereafter condolences poured in. Michael J. Cunningham, president of the Bank of Bisbee, and Lemuel's friend for forty-five years, wrote, "The community has lost a high class, honorable, able man who has done a great deal of good for his fellow man."

Attorney Fred Sutter remarked that Lemuel Shattuck "was an outstanding man, a great character, whom everyone respected and admired."

"He was a man of big character and of high attainments," commented Judge John Wilson Ross. "A friend of the people and of hundreds of them a benefactor."

In the passing of Mr. Shattuck," observed J. G. Flynn, general manager of the Shattuck Denn, "thousands of people have lost a great friend. . . . He was of the type that really made the west and made it great."

"Lem Shattuck was an exceptionally able man and a man who did much for his community and state," wrote pioneer businessman man and rancher, James E. Brophy. "The people of this community will miss him terribly."

On September 10 the Bisbee *Daily Review* published its farewell:

The state has lost an outstanding financier and a man who did not hesitate to back his faith in Arizona's future with his hard earned dollars. Cochise County has lost one of its foremost citizens. The Bisbee district has lost one of its few who took poor years with the good, helping where help was needed, but always with an abiding faith in Bisbee. There are only a few Shattucks in every generation and the people of Bisbee are fortunate because he called Bisbee his home.

The pioneers are passing, those hardy people who made the West. Those of us who come after to enjoy the civilization made possible by their labor do not take sufficient thought of the blessings which they prepared. One by one they leave us and it is doubtful whether those of us remaining have the same fearlessness, the same courage, the same faith which characterized the pioneers of the Southwest. We shall miss him.

A man's accomplishments are usually forgotten within a few years of his death. This is not the case with Lemuel Shattuck. Throughout the depression he supported dozens of old miners, prospectors, and cattlemen, sending them monthly checks with which to buy grub, or pay rent. Gruff, straight talking, frank and outspoken, those in need learned to know his heart was close to the surface. He aided both the little man as well as the powerful. While his acts of generosity are remembered to this day, it was mining ventures and a financial institution that etches Lemuel Shattuck's name to the historical record.

The Shattuck Mine produced for two decades, and paid over $8 million dollars in dividends. When it was depleted, he turned his attention to the Denn Mine, and with the same faith and determination guided its development. The mine had proven ore reserves when the depression struck. Revived in 1934, Shattuck gradually increased the Denn's production, and it is unfortunate that Lemuel did not live to see its 1940 output of 13 million pounds of copper, 3 million pounds of zinc, 318,158 ounces of silver, and 9,446 ounces of gold.[61]

Production meant profit, which allowed the Shattuck Denn Mining Corporation to acquire the Iron King lead-zinc mine near Humboldt, Arizona, and the controlling interest in the 79 Mine, near Hayden Junction, which Lemuel had invested in during the 1920s. The company also purchased the Zuni Milling Company, a fluorspar producer near Grant, New Mexico.

World War II revived the Shattuck Mine. Between 1943 and 1947 leasers extracted 400 tons of copper a month from the old workings.[62] Decline of metal prices again closed it and the Denn. In March of 1947 Phelps Dodge entered into negotiations with Shattuck Denn management to buy the Denn, the price being $300,000 in cash, in addition to royalty on ore recovered in excess of the first 100,000 tons. Shattuck Denn stockholders approved the deal, and the Denn became part of Phelps Dodge Corporation. In 1973 Phelps Dodge acquired the old Shattuck property, erected a head-frame moved from a another mine in the district, repaired the shaft, and installed a hoist. The mine produced minor amounts of copper from the 800 foot level until 1976, when all underground operations ceased at Bisbee.

For fifteen years Bisbee's mines have been closed. One should say in hiatus, for there is still vast low-grade reserves of copper in the Warren District. If advanced extraction techniques and the right commodity price should ever combine to pump life into the old camp, it will likely happen on Shattuck mining property.

The 1907 failure of the Cochise Copper Company, largely owned by Joseph Muheim and Lemuel Shattuck, in no way meant that its fifteen claims in Dubacher Canyon were barren of ore. Copper was there, disseminated and low grade. Knowing that technological advancement would someday unlock the treasure stored on the 172 acres, Lemuel held on to his share of the patented Cochise claims, and emphatically urged the Muheims to do the same. His wisdom paid off.

In the 1970s Occidental Minerals Company, a subsidiary of Occidental Petroleum, leased the property. Their tests verified the existence of considerable tonnage of copper mineralization, too low grade to mine profitably. The lease lapsed. In the meantime, Phelps Dodge Corporation was experimenting with dump leaching and recovery of copper by means of solvent extraction-electrowinning (SX-EW). At a small SX-EW pilot operation at Tyrone, New Mexico, the company produced metal at thirty cents a pound. Similar plants were erected at Morenci and Chino. In August 1987, Phelps Dodge announced plans to explore "a potentially significant copper deposit" at Bisbee—the fifteen claims comprising the Cochise group, long held by the Muheim and Shattuck families.

Under a lease-option agreement, which could ultimately spell a $3 million purchase price, as well as royalties amounting to $1.75

million, Phelps Dodge went to work drilling the property. Tests conducted over a three year period revealed a 150 million ton sulphide ore deposit, with an average grade of 0.49 percent copper. The company formulated the Cochise Project, an operation employing 250 people, that envisioned recovering 90 million pounds of electrowon copper annually. There is every possibility the $80 to $90 million Cochise Project will signal the return of mining to Bisbee.[63]

The Miners and Merchants Bank was just as enduring as Lemuel's mining properties, due to the guidance of Spencer Shattuck. While the Yuma branch was sold to the First National Bank in 1937 for $62,500, the Miners and Merchants Bank continued to serve the Warren District. Like his father, Spencer had faith in the financial institution and in Arizona. He opened a branch at Fort Huachuca in July 1942, which ran until 1947. The Benson branch office was established in 1946, and a branch in Lowell followed the next year. The Fort Huachuca branch was re-established in 1951, as was the Warren office. By 1952 assets of the Miners and Merchants Bank totaled more than $9 million.[64] In 1956 First National Bank purchased all the branches of the Miners and Merchants Bank, and Spencer Shattuck became a board member of the consolidated bank. Thus Lemuel Shattuck's financial institution continues to this day under the revised name of First Interstate Bank of Arizona.

Last but not least, Lemuel left his family a legacy: an estate of cash, stocks, bonds, patented mining claims, and property that included two ranches in Cochise County and farm land in Yuma. In all, $2,773,655 worth of assets. With characteristic foresight, Lemuel willed his estate be set aside as a trust for his heirs. Administered by Spencer until his untimely death in 1960 at age 52, and by Chase Manhattan Bank thereafter, this huge estate benefitted not only his wife, Mary O., and children, but Cochise County, and Arizona as well.

For forty years the trust provided for the Shattuck children: for Mark until his death in December 1942; for Spencer; for Warner, who died in 1964; and for Dorothy, who died of cancer a year later.

The legacy also assured a worry-free life for Mary O., who passed away in Warrensburg, Missouri, in 1989, at age ninety-seven. Her son, Lemuel, Jr., grew to manhood in Warren, attended local schools, and graduated from the University of Arizona with a degree in business and public administration. In 1952 he was commissioned a lieutenant in the U.S. Air Force, in which he served three years. Following military service, Lem, Jr., managed the O.K. Ranch for a number of years, and in 1956 sought the Republican nomination for the Arizona House of Representatives for District Nine. He lives today in Tucson, a supporter of civic projects.

Isabel Fathauer, Lemuel's only living child by his first marriage, carries on the Shattuck tradition of ranching and community service. In 1941 she and her husband Walter purchased the IV Bar Ranch in the far southeast corner of Arizona, near the border with New Mexico. Under Spencer's tutelage, Walter Fathauer became a rancher, a far cry from designing x-ray equipment for General Electric. He assisted Spencer in the management of all the Shattuck cattle enterprises. In 1946 Isabel and Walter moved to Tucson to raise their daughters Mari and Dorothy, and devote time and energy to community and University projects, as well as Republican party activities.

Lastly, the cattleman tradition, which originated with the Erie Cattle Company, still burns within the Shattuck family. Although the Bar Boot Ranch and the Yuma Valley farms were sold years ago to pay inheritance taxes, the O.K. Ranch became solely a Shattuck enterprise when Spencer purchased William Lutley's interest. It and the IV Bar spread remain in the family today, managed by the Fathauer's grandson, Walter Lane.

NOTES TO THE CHAPTERS

Chapter One

1. Genealogical data regarding the Shattuck family is found in Lemuel Shattuck (comp.), *Memorials of the Descendants of William Shattuck, the Progenitor of the Families in America that have Borne His Name.* Boston: Dutton and Wentworth, 1855.

2. For genealogical data relative to the Coover family see Melanchthon Coover (comp.), *A Limited Genealogy of the Kober-Cover-Coover Family and Cognate Families.* Gettysburg, Pennsylvania: Melanchthon Coover, 1942.

3. Thomas Baldwin and J. Thomas. *A New and Complete Gazetteer of the United States.* Philadelphia: Lippincott, Grambo & Company, 1854, pp. 362-263.

4. Data relative to the activities of Enoch and Jonas Shattuck in Kansas and the Cherokee Strip was provided by Dan Shattuck, owner of the present-day Shattuck Ranch at Ashland, Kansas, and grandson to Enoch. Much of this material was compiled from the Medicine Lodge *Cresset* and the Barber County *Mail,* from May 21, 1878 to December 6, 1883.

5. Additional details of the range cattle business along the Cherokee Strip is found in William W. Savage, Jr. *The Cherokee Strip Live Stock Association: Federal Regulation and the Cattleman's Last Frontier.* Columbia: University of Missouri Press, 1973.

6. Allen A. Erwin. *The Southwest of John H. Slaughter, 1841–1922.* Glendale: The Arthur H. Clark Company, 1965.

7. Originally this vast grassland was labeled the Sulphur Spring Valley. Usage over the years has made springs plural. See O. E. Meinzer and F. C. Kelton. *The Geology and Water Resources of Sulphur Spring Valley, Arizona.* U.S.G.S. Water Supply Paper No. 320. Washington, D.C.: Government Printing Office, 1913.

8. Incorporation papers for the Erie Cattle Company on file in Cochise County Court House, Bisbee, Arizona.

9. For the process of pre-emption see Paul W. Gates. *History of Public Land Law Development.* Washington, D.C.: Government Printing Office, 1968. See also Thomas Donaldson. *The Public Domain: Its History with Statistics.* New York: Johnson Reprint Corp., 1970.

10. The land and water rights holdings of the Erie Cattle Company, and principals starting at the Mexican border and working north and east:

H. H. Whitney land in Sec 10 & 15–T24S, R27E–eleven miles west of Douglas, Arizona, town site 160 acres, Whitewater Draw. 160 acres.

H. H. Whitney land Sec 28 & 29, T23S, R27E on Whitewater Draw, three miles north and four miles west of site of Douglas. 160 acres.

J. H. Shattuck possession rights in Sec 12–T23S, R2, 6E, Whitewater Draw. 160 acres.

E. A. Shattuck Sec 4, T23S, R26E–Whitewater Draw, near site of Double Adobe, twelve miles east of Lowell. 160 acres.

Erie Cattle Company "Asa Turner Ranch," Sec 20 & 21, T22S, R26E. Whitewater Draw, two miles north, one mile west of Double Adobe. 160 acres.

W. W. Whitney land NW¼NE¼ Sec 15, T22S, R27E, Mud Springs Draw, one mile south of Taylor Buttes. 40 acres.

Erie Cattle Company land NW¼ Sec 21, T20S, R27E, five miles east of Elfrida, sold to William Lutley in 1890, "EM Bar B" Ranch. 160 acres.

Erie Cattle Company water appropriation Mud Springs, about four miles northwest of College Peak. 1894–present Mud Springs Ranch, Sec 22, T22S, R28E. 160 acres.

Erie Cattle Company water appropriation, stream, spring and wash, and reservoir site, situated near Silver Creek Ranch in Silver Creek Draw about two miles east of College Peak. Sec 4, T23S, R29E or Sec 33, T22S, R29E. 160 acres.

J. E. McNair SW¼SW¼ Sec 34, T20S, R29E, sold to William Fuller in 1900. Northern edge of present-day Coronado National Forest, southwest of Limestone Peak, head of High Lonesome Canyon. 40 acres.

Erie Cattle Company water appropriation, reservoir, construction situated in southern end of Chiricahua Mountains in what is known as Texas Canyon, known as "The Tanks." 160 acres.

J. E. McNair, Chiricahua Mountains possession rights W½W½ Sec 19, T21S, R30E, on Buck Creek, near present east edge of Coronado National Forest (present Glenn Ranch). 80 acres.

Leslie Ranch, Swisshelms Sec 21, T21S, R28E, mouth of High Lonesome Canyon 1896. 160 acres.

Mulberry Ranch, S½SW Sec 5, T22S, R26E, Whitewater Draw. 160 acres, acquired 1896.

Linderman & Banks Ranches (northern Swisshelms. 320 acres, acquired 1897.

Total estimated holdings in acres of Erie Cattle Company and principals–2240 acres.

11. A contemporary account of ranching in the Sulphur Springs Valley is found in Joseph A. Munk. *Arizona Sketches*. New York: Grafton Press, 1905.

12. Bisbee *Orb*, July 3, 1898.

13. In an interview given to Mary D. McFadden, Lemuel Shattuck stated he "fought Indians with Colonel Greene." This statement corroborates many claims of Colonel William C. Greene having taken the field against Apache Indians. Although Shattuck always claimed he fought Apaches, this is the only time he definitely states with whom he campaigned. See C. L. Sonnichsen. *Colonel Greene and the Copper Skyrocket*. Tucson: University of Arizona Press, 1974, pp. 21–23. Mary D. McFadden's interview, published in a St. Paul, Minnesota newspaper, was reprinted in the Bisbee *Daily Review*, July 17, 1910, second section, p. 3.

14. Anna Belle Bootes of Waterford, Pennsylvania, graduated from the Nor-

mal School at Edinboro, Pennsylvania, and taught school before her marriage in 1885. She was an artist of considerable talent. Her oil paintings of Arizona wildflowers are a cherished possession of her descendants.

15. Overlock biographical files, Arizona Historical Society, Tucson, Arizona.

16. Joe Chisholm. *Brewery Gulch: Frontier Days of Old Arizona — Last Outpost of the Great Southwest.* San Antonio: The Naylor Company, 1949, pp. 146–149.

Chapter Two

1. Odie B. Faulk. *Tombstone: Myth and Reality.* New York: Oxford University Press, 1972.

2. Lynn R. Bailey. *Bisbee: Queen of the Copper Camps.* Tucson: Westernlore Press, 1983.

3. According to old records Mule Gulch or Canyon was the proper name for the defile in which Bisbee's main street would develop. Tombstone Canyon was the designation of the gorge on the eastern side of the divide.

4. Bailey, pp. 33–34, 49.

5. Loraine Mackintosh. "The Day the Valley Shook," *The Cochise Quarterly*, Vol. 14, No. 2, Summer 1984, pp. 3–6.

6. L. C. Shattuck to Thomas Bardon, May 9, 1917. Fathauer Collection. Hereafter cited by correspondents and date.

7. Ibid. The Vulture Mine, discovered in 1863 by Henry Wickenburg, produced over 16 million dollars in gold. In 1882 an 80-stamp mill was installed. See Bisbee *Daily Review*, September 20, 1911, p. 4, for additional details relative to the Vulture Mine.

8. For detailed discussion of the Arizona and South Eastern Rail Road Company see: David F. Myrick, *Railroads of Arizona.* Berkeley: Howell-North Books, 1975, Vol. 1.

9. L. C. Shattuck to Chief of U.S. Immigration Service, June 4, 1927.

10. Tombstone *Epitaph*, February 8, 1890.

11. Tombstone *Prospector*, March 17, 1891.

12. Ibid., April 19, 1892.

13. It has been asserted that L. C. Shattuck's Lumber Company was Bisbee's first such enterprise. That was not so. George Preston had a lumber company in Bisbee during the 1880s, which sold lumber obtained from mills in the Chiricahua and Huachuca mountains, and the Copper Queen Mercantile Company dealt in lumber as well. It is not known precisely when Shattuck entered this trade. The filing of liens during summer of 1892 for payment of building materials suggests that it must have been sometime between January and June 1892. See filing of lien by L. C. Shattuck on house of James Campbell, Bisbee, to secure sum of $149.36. Published in Tombstone *Prospector*, June 28, 1892; also Shattuck vs. William Liggett, *Prospector*, October 7, 1893; and Shattuck vs. E. G. Norton, *Prospector*, November 27, 1893.

14. Tombstone *Prospector*, March 3, 1894.

15. Ibid., November 17, 1894.

16. Ibid., November 17, 1894.

17. An 1899 ledger for the Bisbee Beer Bottling Company records 216 accounts including the Anheuser Saloon, the Copper Queen Store, Baptiste Caretto, the Bisbee Band, J. B. Angius, the Brewery Gulch Restaurant, Cafe Royal, the Elks, John Henninger (Tombstone), William King (Tombstone), Letson & Company, Joseph Muheim, Medigovich & Nobile, Miners Saloon, Opera Club Saloon, Palace Saloon, Bob Tate, Turf Saloon, Dan Walsh, and a Miss Clara. Account book of Bisbee Beer Bottling Company, July 1899, in Fathauer Collection.

18. For additional details relative to Bisbee lumber companies see Tombstone *Prospector*, March 4, 1896; January 1, 18, March 18, 1899.

19. Development of the World's Fair claims is covered in a lengthy article in Bisbee *Daily Review*, January 25, 1903; February 6, 1903.

20. Tombstone *Prospector*, May 27, September 11, 1900.

21. Bisbee *Daily Review*, April 29, 1902.

22. Bisbee *Daily Review*, September 7, 1902.

Chapter Three

1. Bisbee *Daily Review*, September 7, 1902, p. 1.

2. Ibid., December 27, 1901, p. 1.

3. Ibid., December 14, 1901, p. 1.

4. Tombstone *Prospector*, April 31, 1894. Lemuel Shattuck's name appears in the *Prospector* of September 23, 1892, on a list of twenty-five possible delegates to the County Democratic Convention. Sixteen were picked, including Shattuck. He was an alternate candidate to the Democratic Territorial Convention held in Phoenix September 17, 1894. (Ibid., September 11, 1894.) In September 1896 Lemuel's name was put forward as a possible delegate to the County Democratic Convention. (Ibid., September 22, 1896.)

5. Bailey, *Bisbee: Queen of the Copper Camps*, pp. 92–93.

6. Bisbee *Daily Review*, August 19, 1903, p. 1.

7. Actually L. O. Milless was born in Baltimore, Maryland, 1851. He moved with his father to Des Moines, Iowa, and later to Carthage, Missouri. (Ibid., August 20, September 20, 1903.) Milless's probate records in the Cochise County Courthouse reveal he had a home in Douglas.

8. Bisbee *Daily Review*, August 26, 1903.

9. Ibid., September 2, 1903.

10. Ibid., October 1, 2, 7, 1903.

11. Lemuel Shattuck and E. B. Mason built the Opera House in winter of 1900. (Tombstone *Prospector*, November 26, 1900.)

12. Bisbee *Daily Review*, October 1, 1903.

13. Noftz's letter was published in ibid., October 29, 1903.

14. Ibid., May 3, 1904.

15. Ibid., May 5, 6, 7, 1904.

16. Ibid., May 19, 1906.

17. Ibid., October 29, 1908, p. 5; also Douglas *Daily International*, October 25, 1908, p. 3; October 29, 1908, p. 2.

18. Bailey, *Bisbee: Queen of the Copper Camps*, p. 82.

19. Bisbee *Daily Review*, June 14, 18, 1902.
20. Ibid., July 17, 1902.
21. Ibid., July 30, 1902.
22. Ibid., June 5, 1903.
23. Ibid., March 23, 1904.

Chapter Four

1. Benjamin F. Graham showed up in Arizona about 1882. Like Shattuck, he worked as a cowboy for various Sulphur Springs Valley ranchers, and drifted into Bisbee. In 1898 he started a livery stable, and shortly after incorporation of the town served as its first commissioner of streets, furnishing a buckboard for the pesthouse and a team for the scavenger wagon (See Minutes of the Bisbee City Council, February 4, 1902, p. 7). Graham also founded the Bisbee-Naco Stage Line (Bisbee *Daily Review*, February 8, 1902, p. 3), and developed mining properties under the name B. F. Graham and Company, with offices in the Copper Queen Hotel. (Bisbee *Daily Review*, February 27, 1903, p. 2.)

2. For a detailed discussion of El Tigre see *Mines and Minerals*, Vol. XXIX, No. 11, June 1909, pp. 483–485.

3. Although this dialogue is reconstructed, Shattuck often related his meeting with Graham to family members. See ibid., as well as Bisbee *Daily Review*, March 9, 1903, p. 2.

4. *Mines and Minerals*, pp. 485-486.

5. Bisbee *Daily Review*, February 17, 1903, p. 3.

6. Ibid., February 22, 1903, p. 4.

7. Ibid., February 6, 1903, p. 7.

8. Douglas *Daily International-American*, January 18, 1906, p. 1.

9. Frank Aley detailed this six-day trip in the Bisbee *Daily Review*, March 6, 7, 8, 1903.

10. Ibid., March 9, 1903, p. 7.

11. Ibid.

12. Ibid., March 25, 1904, p. 4; May 21, 1905, p. 5.

13. Ibid., March 21, 1903, p. 8.

14. Ibid., May 8, 1903, p. 5.

15. Ibid., May 25, 1903, p. 5.

16. Ibid., February 9, 1905, p. 2; see Lucky Tiger Annual Report published in ibid., March 12, 1905, p. 8.

17. Ibid., July 31, 1907, p. 2.

18. Douglas *Daily International-American*, July 31, 1905, p. 1.

19. Bisbee *Daily Review*, October 11, 1905, pp. 2 and 3.

20. Ibid., August 30, 1905, pp. 1 and 8; January 19, 1906, p. 2; also *Mines and Minerals*, op. cit. See also L. C. Shattuck to Porfirio Diaz, December 6, 1905, Fathauer Collection.

21. Douglas *Daily International-American*, July 17, 1905, p. 1.

22. Ibid., July 31, 1905, p. 1.

23. Ibid., July 17, 1905, p. 1.

24. Ibid.

25. Bisbee *Daily Review*, December 31, 1905, p. 1.

26. Ibid., July 8, 1905, p. 1.

27. Ibid., September, 10, 1905, p. 7.

28. Douglas *Daily International-American*, July 17, 1905, p. 1.

29. Ibid., August 7, 1905, p. 1.

30. Ibid., August 11, 1905, p. 2.

31. Bisbee *Daily Review*, December 15, 1905, p. 1.

32. Douglas *Daily International-American*, August 29, 1905, p. 2; Bisbee *Daily Review*, September 14, 1905, p. 3.

33. Douglas *Daily International-American*, September 19, 1905, p. 1; Bisbee *Daily Review*, October 10, 1905, p. 1. See also L. C. Shattuck to Porfirio Diaz, December 6, 1905, op. cit.

34. Douglas *Daily International-American*, September 16, 1905, p. 1.

35. Bisbee *Daily Review*, January 30, 1909, p. 5. B. F. Graham, John Brooks and T. J. Wiley were ordered held for robbery and despoilation. They did not report for trial, forfeited their bonds, and became fugitives from Mexican justice.

On February 17, 1906, Graham resigned as president of the Ensenada Mining Company. His place was taken by Sol Epes Randolph of Tucson, Arizona, personal representative of Edward H. Harriman of the Southern Pacific Railroad Company. Randolph directed the remaining litigation on behalf of the Ensenada Company against the Lucky Tiger Company. The participation of Sol Epes Randolph indicates that El Tigre would have gone to the Southern Pacific Company if Graham had won his case.

Graham's flight to British Columbia in no way deterred him. In fall of 1906 he purchased 140,000 acres of Canadian timberland, and established the Graham Island Lumber Company and the Graham Steamship, Coal and Lumber Company, which shipped lumber throughout the Pacific Northwest. He later sold 114,000 acres to eastern capitalists, and as a millionaire visited Bisbee in late February 1909. Graham settled his family permanently in Southern California, where he had considerable property. (Bisbee *Daily Review*, February 20, 1909, p. 5.)

36. O. E. McMuller to L. C. Shattuck, February 5, 1929; L. C. Shattuck to O. E. McMuller, February 9, 1929. Shattuck's Lucky Tiger stock eventually went to his youngest daughter, Dorothy I. Shattuck, as part of Isabella Shattuck's estate.

37. Bisbee *Daily Review*, August 13, 1908, p. 6.

38. Ibid., October 31, 1915, p. 1.

39. Ibid., June 28, 1917, p. 1.

40. In December 1913 El Tigre broke a production record by producing seventy-three bars of bullion in one month. (Ibid., January 9, 1914, p. 3.) By January 1914 the mine had entirely retired its indebtedness. (Ibid., January 28, 1914, p. 2.)

Chapter Five

1. Development of the Irish Mag Mine and the organization of the Calumet and Arizona Mining Company are covered in the following sources: Richard W. Graeme, "Bisbee, Arizona," *The Mineralogical Record*, Vol. XII, No. 5, September-October 1981, pp. 260–279; Bailey, *Bisbee: Queen of the Copper Camps*,

pp. 40–45; and Robert Glass Cleland, *A History of Phelps Dodge, 1834–1950.*
New York: Alfred A. Knopf, 1952, pp. 119–123.

2. The Bisbee *Daily Review*, February 24, 1903, lists the World's Fair group among the most promising undeveloped claims in the Warren District, along with the Copper Glance, the Modern, and the Cochise.

3. Ibid., July 3, 1903, p. 4.

4. Ibid., May 27, 1903, p. 6.

5. Ibid., July 3, 1903, p. 6.

6. Ibid., October 3, 1903, p. 6.

7. Ibid., April 1, 1904, p. 4.

8. Ibid., August 21, 1904, p. 5.

9. Ibid., February 19, 1905, p. 10.

10. Ibid., March 8, 1905, p. 2; May 30, 1905, p. 5; July 29, 1906, p. 5.

11. A lengthy article in ibid., May 7, 1904, p. 3, details the advent of Michigan mining men on the Bisbee scene.

12. Ibid., January 30, 1902, p. 4.

13. Ibid.

14. Ibid.

15. This was a favorite story of Lemuel's, often related to family and friends.

16. Bisbee *Daily Review*, September 22, 1912, p. 1.

17. Ibid., June 9, 1908, p. 6.

18. Ibid., March 25, 1904, p. 8; May 16, 1906, p. 8.

19. Reprinted in ibid., April 7, 1904, p. 3.

20. Ibid., September 22, 1904, p. 6.

21. The claims constituting the Denn Development Company were the Triangle, Robert E. Lee, Cotton Tail, York State, Orleans, Bucky O'Neill, Xenis, Examiner, Ophir, Williams, Buzzard, Admiral Dewey, and part of the John Daisy. Ibid., February 19, 1905, p. 10; December 14, 1906, p. 6.

22. Ibid., February 2, 1905, p. 8.

23. Ibid., March 5, 1905, p. 10.

24. Ibid., July 26, 1905, p. 10.

25. Ibid., October 10, 1905, p. 2; Oct. 11, 1905, p. 2.

26. Ibid., November 26, 1905, pp. 1 and 9.

27. L. C. Shattuck to F. H. Brainerd, August 5, 1908. Fathauer Collection. Hereafter cited by correspondents and dates.

28. Bisbee *Daily Review*, August 18, 1906, p. 8.

29. Graeme, p. 311.

30. Bisbee *Daily Review*, October 16, 1906, p. 5.

31. Ibid., January 17, 1907, p. 7.

32. The officers of the Denn-Arizona Development Company were: Martin Pattison, president; Thomas Bardon, vice president; L. C. Shattuck, treasurer; John G. Williams, secretary. A. Guthrie, A. M. Chisholm, John G. Williams, Thomas Bardon, Byron M. Pattison, Martin Pattison, Maurice Denn, and L. C. Shattuck were directors. Ibid., March 2, 1905, pp. 5 and 8; December 14, 1906, p. 6.

33. Ibid., December 1, 1904, p. 2; February 19, 1905, p. 1.

34. Ibid., March 5, 1905, p. 10.

35. Ibid., August 4, 1907, p. 6.

36. Ibid., December 11, 1906, p. 6; December 14, 1906, p. 6.

37. Ibid., January 17, 1907, p. 8.

38. The claims comprising the Cochise group were: Wee Wee, Red Hill, Carolina, Paragon, Key to the Situation, Leviathan, Henry George, Nancy Hanks, Jack of Clubs, La Luisa, McGinty, Yo Tambien, Sulphide, Pajaro, and Hill. Directors Report to Stockholders, attached to Minutes of the Cochise Development Company, Fathauer Collection. Hereafter cited as Minutes. See also Bisbee *Daily Review*, October 1, 1905, p. 9.

39. Minutes of Special Stockholders' Meeting of April 20, 1914, appended to Minutes of Cochise Development Company, p. 57. See also L. C. Shattuck to John C. Van Clef, January 10, 1906.

40. Bisbee *Daily Review*, October 1, 1905, p. 9; Directors Report to Stockholders, January 8, 1906, appended to Minutes.

41. *Daily Review*, May 6, 1906, p. 1.

42. Ibid., September 4, 1906; Directors Report, January 14, 1908, Minutes.

43. Minutes of Special Meeting of the Directors, April 1, 1914, p. 55–56; Minutes of Directors, May 9, 1917; all in ibid.

Chapter Six

1. Tombstone *Prospector*, November 19, 1900, p. 2.

2. Bisbee *Daily Review*, September 11, 1902, p. 2.

3. For additional details relative to the history of Douglas consult Robert S. Jeffrey, "The History of Douglas, Arizona." Unpublished Master's thesis, University of Arizona, 1951.

4. Tombstone *Epitaph*, May 26, 1901, p. 3; *Prospector*, June 30, 1901, p. 2.

5. Shattuck also owned residential property in Douglas at the corner of 12th and D streets. Bisbee *Daily Review*, November 8, 1911, p. 3.

6. Ibid., June 23, 1906, p. 8; June 29, 1906, p. 4; July 19, 1906, p. 5.

7. Ibid., May 6, 1903, p. 4.

8. Ibid., June 27, 1903, p. 2; July 17, 1905, p. 4.

9. Ibid.

10. Ibid., May 9, 1904, p. 5; February 24, 1905, p. 6.

11. Ibid., March 23, 1904, p. 2; August 5, 1904, p. 5.

12. Ibid., October 23, 1904, p. 6.

13. Ibid., July 21, 1905, p. 6.

14. Ibid., May 17, 1906, p. 4; April 4, 1907, p. 5.

15. Ibid., July 24, 1908, p. 5; July 26, 1908, p. 5.

16. Rawle B. Coover, of Geary, Pennsylvania, was assistant cashier of the Miners and Merchants Bank. Ibid., August 29, 1914, p. 6.

17. Ibid., March 15, 1907, p. 8; August 18, 1907, p. 1.

18. Ibid., November 5, 1907, p. 1.

19. Ibid., November 8, 1907, p. 5; November 12, 1907, p. 5.

20. Ibid., September 8, 1907, p. 4; September 19, 1907, p. 5; September 21, 1907, p. 1.

21. Harry Duey, secretary of the Cochise Development Company, collected calls amounting to $3,000 from stockholders, and instead of depositing them to

the credit of the company, turned them over to the brokerage firm of Duey and Overlock. Duey was voted off the company's board of directors, and prosecuted by Shattuck. See Minutes of the Cochise Copper Company, October 23, 1907, p. 24. Fathauer Collection.

22. Bisbee *Daily Review*, October 6, 1907, p. 5.

23. J. H. Nolan and W. J. Eddleman, officers of the First National Bank of Bisbee, were sentenced to five years' imprisonment for embezzlement. Ibid., May 29, 1909, p. 1.

24. Ibid., August 18, 1907, p. 1.

25. Ibid., November 5, 1907, p. 1; November 7, 1907, p. 1.

26. Ibid., November 16, 1907, p. 1; November 17, 1907, p. 1.

27. Ibid., January 28, 1910, p. 7.

28. Thomas Bardon to L. C. Shattuck, March 28, 1910. Fathauer Collection. Hereafter cited by correspondents and dates.

29. Ibid., March 29, 1910.

30. L. C. Shattuck to Fred Merritt, March 9, 1910.

31. Ibid.

32. Fred Merritt to L. C. Shattuck, April 27, 1910.

33. Telegram, L. C. Shattuck to Fred Merritt, May 8, 1910.

34. *Engineering and Mining Journal*, Vol. 91, May 13, 1911, p. 979; History of Denn Mine, Fathauer Collection.

35. L. C. Shattuck to Thomas Bardon, November 20, 1907; *Engineering and Mining Journal*, Vol. 85, No. 4, January 25, 1908, p. 197.

36. *Engineering and Mining Journal*, Vol. 86, No. 8, August 22, 1908, p. 39.

37. Bisbee *Daily Review*, May 25, 1909, p. 6.

38. Ibid., December 5, 1908, p. 1.

39. *Engineering and Mining Journal*, Vol. 90, No. 15, October 8, 1910, p. 733.

40. Bisbee *Daily Review*, February 5, 1909, p. 5; October 8, 1910, p. 733.

41. L. C. Shattuck to Martin Pattison, April 27, 1909; *Engineering and Mining Journal*, Vol. 91, May 13, 1911, p. 979.

42. *Engineering and Mining Journal*, Vol. 89, No. 16, April 16, 1910, p. 839.

43. Ibid., Vol. 90, No. 15, October 8, 1910, p. 733.

44. Bisbee *Daily Review*, April 3, 1910, p. 6.

45. Shattuck presented Weed and Probert with 100 shares of Shattuck-Arizona stock. L. C. Shattuck to Thomas Bardon, May 20, 1910; Weed and Probert to L. C. Shattuck, July 26, 1910.

46. Frank H. Probert to Galen L. Stone, April 28, 1910.

47. L. C. Shattuck to Directors of Shattuck-Arizona Mining Company, June 6, 1907; Shattuck to Thomas Bardon, June 1, 1907.

48. L. C. Shattuck to Thomas Bardon, April 1, 1907.

49. Bisbee *Daily Review*, June 16, 1912, p. 4.

50. Reprinted in ibid., November 24, 1912, p. 2.

51. Ibid., January 19, 1913, p. 1. Like other caves in the Warren District, the Shattuck cavern was created by oxidation of ore bodies. Circular in shape, with a diameter of 340 feet, and a height of 172 feet, the cavity was the largest encountered at Bisbee. Breathtakingly beautiful — with green, blue, and brown calcite and aragonite formations — the cave was left open for public visitation for many

weeks. Ibid., April 20, 1913, p. 2. For detailed discussion of the cavern see Philip D. Wilson, "A Cavern in the Shattuck Mine," *Engineering and Mining Journal*, Vol. 97, No. 15, April 11, 1913, pp. 743-44.

52. L. C. Shattuck to Anheuser-Busch, May 16, June 4, 1908.

53. Shattuck's membership in the Knights of the Royal Arch is attested to by donations he made to the organization. L. C. Shattuck to Anheuser-Busch, April 14, 27, 1909.

54. Bisbee *Daily Review*, October 28, 1909, p. 1; March 12, 1910, p. 8.

55. Ibid., November 3, 1909, p. 1.

56. L. C. Shattuck to Anheuser-Busch, July 2, 1914.

57. There is no doubting this statement on the basis of the amount of wholesale beer purchased by Shattuck from Anheuser-Busch. Lemuel's Bisbee distributorship bought from $5,000 to $8,000 worth of keg beer monthly. L. C. Shattuck to Anheuser-Busch, October 5, 1908, June 13, 1908, and February 2, 1909.

58. Bisbee *Daily Review*, March 24, 1910, p. 6.

59. The various holdings of the Erie Cattle Company were purchased in fall of 1900 by the Ryan Brothers Cattle Company, which had extensive ranches in Montana and Kansas. Tombstone *Prospector*, September 8, 1900, p. 4.

60. Burton Mossman to L. C. Shattuck, April 24, 1908; Charles J. Hysham to L. C. Shattuck, October 30, 1908.

61. Tombstone *Prospector*, April 5, 1914, p. 3.

Chapter Seven

1. Bisbee *Daily Review*, April 19, 20, 1911.

2. L. C. Shattuck to Whom It May Concern, August 31, 1907.

3. L. C. Shattuck to B. Jazdzewski, November 5, 1922.

4. L. C. Shattuck to Sister Catherine, May 3, 1907.

5. Ibid., June 16, 1908.

6. James Wadham to L. C. Shattuck, February 24, 1909; L. C. Shattuck to James Wadham, March 29, 1909.

7. L. C. Shattuck to Daisy Grenfell, June 15, 1910.

8. James Wadham to L. C. Shattuck, November 4, 1914; L. C. Shattuck to James Wadham, September 28, 1914.

9. James Wadham to L. C. Shattuck, October 1, 1914.

10. Rufus Grenfell to L. C. Shattuck, June 10, 1916.

11. L. C. Shattuck to Rufus Grenfell, June 18, 1916.

12. Charles Pritchard to L. C. Shattuck, October 28, 1916.

13. L. C. Shattuck to James Wadham, December 28, 1916.

Chapter Eight

1. Bisbee *Daily Review*, February 8, 1914, p. 3,

2. Ibid., February 26, 1914, p. 3.

3. Ibid., September 29, 1914, p. 8.

4. Ibid., July 7, 1915, p. 1; July 8, p. 1.

5. Ibid., September 14, 1917, p. 5.

6. Ibid., April 6, 1917, p. 5.

7. Ibid., May 19, 1918, p. 1.

8. Patrick Renshaw, *The Wobblies: The Story of Syndicalism in the United States.* New York: Doubleday & Company, Inc., 1967.

9. Thomas Campbell to L. C. Shattuck, April 10 and 17, 1917; L. C. Shattuck to Thomas Campbell, April 10, 1917.

10. Thomas Bardon to L. C. Shattuck, April 5 and 7, 1917.

11. L. C. Shattuck to Thomas Bardon, June 27, 1917; Thomas Bardon to L. C. Shattuck, June 27, 1917.

12. L. C. Shattuck to Thomas Bardon, June 27, 1917.

13. Ibid., June 27, 1917.

14. Bisbee *Daily Review*, June 27, 1917, p. 3.

15. Ibid., June 27, 1917, p. 1.

16. P. Buckwalter to L. C. Shattuck, August 12, 1916.

17. Certificate of Deputization, July 3, 1917; in Fathauer Collection.

18. Testimony of Frank A. Thomas, in U.S. Committee on Public Information. "Report of President Wilson's Mediation Commission on the Bisbee, Arizona, Deportations, Including its Recommendations as Made on Findings of Facts," *Official Bulletin*, Vol. I, No. 170, November 27, 1917, p. 147.

19. Thomas Bardon to L. C. Shattuck, July 6, 1917.

20. *Dunbar's Weekly*, Vol. IV, No. 37, September 22, 1917, p. 1.

21. L. C. Shattuck to E. B. Perrin, July 17, 1917.

22. L. C. Shattuck to Thomas Bardon, August 29, 1919.

23. L. C. Shattuck to Thomas Bardon, June 17, 1922; Gordon Campbell to L. C. Shattuck, July 6, 1922; L. C. Shattuck to A. M. Chisholm, June 17, 1922.

24. Bisbee *Daily Review*, August 28, 1917, p. 1.

25. Ibid., September 14, 1917, p. 5.

26. Ibid., September 20, 1917, p. 1.

27. Ibid., September 20, 1917, p. 2.

28. Ibid., September 22, 1917, p. 2.

29. Edward Grover to L. C. Shattuck, June 29, July 16 and 31, August 6; October 14, 1921.

30. L. C. Shattuck to Bureau of Vital Statistics, Trenton, New Jersey, May 5, 1920.

31. L. C. Shattuck to J. E. James, June 12, 1920.

32. L. C. Shattuck to Thomas Bardon, July 3, 1917.

33. L. C. Shattuck to Kathrine Shattuck, December 2, 1920.

34. E. C. Anderson to L. C. Shattuck, December 23, 1920; L. C. Shattuck to E. C. Anderson, December 29, 1920.

Chapter Nine

1. Bisbee *Daily Review*, June 5, 1918, p. 6; *Engineering & Mining Journal*, Vol. 104, No. 7, August 18, 1917, p 322. Hereafter cited as *EMJ*.

2. Bisbee *Daily Review*, February 16, 1918, p. 1.

3. *EMJ*, Vol. 104, No. 6, August 11, 1917, 278; Vol. 104, No. 5, August 4, 1917, 232.

4. *See* Glen L. Allen, "The Shattuck-Arizona Mill for Concentrating Silver Lead-Carbonate Ores," *EMJ*, Vol. 110, No. 16, October 16, 1920, pp.759-762.

5. The contract between J. M. Callow and the Shattuck-Arizona Company specified payment of $3,000 for use of patent rights, and $1,500 per flotation head installed in the mill, plus two-and-a-half cent royalty per ton of concentrate produced. Arthur Houle to L. C. Shattuck, Sept. 9, 1917.

6. Telegram, John G. Williams to L. C. Shattuck, December 20, 1918.

7. L. C. Shattuck to Thomas Bardon, March 14, 1919. Ibid.

8. *EMJ*, Vol. 109, No. 15, April 10, 1920, p. 907.

9. L. C. Shattuck to Byron N. Pattison, June 11, 1921.

10. *EMJ*, Vol. 111, No. 13, March 26, 1921, p. 571.

11. Bisbee *Daily Review*, December 7, 1913, mining section, p. 9.

12. L. C. Shattuck to Mrs. A. G. Watkins, April 17, 1916.

13. *See* Memorandum of Agreement, enclosure Gordon Campbell to Thomas Bardon, March 29, 1917. Ibid.

14. Martin Pattison to L. C. Shattuck, December 2, 1915; L. C. Shattuck to Pattison, December 22, 1915. Ibid.

15. Report of Denn-Arizona stockholders meeting, June 21, 1917; Tomas Bardon to L. C. Shattuck, April 23, 1917; Bisbee *Daily Review*, May 13, 1917, mining section, p. 1.

16. L. C. Shattuck to Thomas Bardon, March 14, 1919.

17. L. C. Shattuck to H. L. Mundy, July 31, 1919.

18. L. C. Shattuck to Thomas Bardon, Feb. 27, 1920.

19. Telegram, Thomas Bardon to L. C. Shattuck, July 18, 1921.

20. L. C. Shattuck to Thomas Bardon, Jr., February 2, 1923.

21. Thomas Bardon, Jr. to L. C. Shattuck, May 7, 1923; Gordon Campbell to L. C. Shattuck, July 20, 1923.

22. *EMJ*, Vol. 117, No. 15, April 12, 1924, p. 630.

23. L. C. Shattuck to Thomas Bardon, Jr., November 3, 1923.

24. *EMJ*, Vol. 119, No. 17, April 25, 1925, p. 697; Vol. 119, No. 22, May 30, 1925, p.897.

25. Charles Kappler to L. C. Shattuck, July 3, 1925.

26. L. C. Shattuck to Charles Kappler, July 17, 1925.

27. *EMJ*, June 20, 1925, p. 1015.

28. *EMJ*, Vol. 120, No. 21, November 21, 1925, p. 827.

29. *EMJ*, Vol. 122, No. 2, July 10, 1926, p. 63.

30. L. C. Shattuck to P. H. Goodhart, May 11, 1928.

31. *EMJ*, Vol. 126, No. 10, September 8, 1928, p. 338; Vol. 126, No. 24, December 15, 1928, p. 956.

32. L. C. Shattuck to Thomas Bardon, Jr., January 8, 1929.

33. W. C. Humphrey to L. C. Shattuck, January 14, 1929.

34. L. C. Shattuck to M. W. Bacon, August 16, 1930.

35. *EMJ*, Vol. 126, No. 9, September 1, 1928, p. 348.

36. L. C. Shattuck to Thomas Bardon, Jr., January 18; August 12, 1929.

37. Telegram, Thomas Bardon, Jr. to L. C. Shattuck, October 29, 1929; L. C. Shattuck to Thomas Bardon, Jr., November 14, 1929.

38. L. C. Shattuck to Thomas Bardon, Jr., June 30, 1930.

39. L. C. Shattuck to Thomas Bardon, Jr., July 1, 1930.

40. L. C. Shattuck to Thomas Bardon, Jr., October 1, 1930.

41. L. C. Shattuck to Thomas Bardon, Jr., September 10, 1930; Thomas Bardon, Jr. to L. C. Shattuck, October 7, 1930.

42. L. C. Shattuck to Thomas Bardon, Jr., October 15, 1930.

43. For additional details relative to deepening and concreting the Denn shaft see: F. P. Brunel, "Concreting the Denn Shaft," *EMJ*, Vol. 133, No. 12, December 1932, pp. 614-617.

Chapter Ten

1. Byron Pattison to L. C. Shattuck, November 28, 1911.

2. L. C. Shattuck to California Experiment Station, Berkeley, California, July 19, 1911.

3. Bisbee *Daily Review*, March 1, 1910. For additional information relative to Yuma Reclamation Project see: *Facts, Figures and Pictures About the City of Yuma and the 150,000 Acres of Valley, and Mesa Land Along the Colorado River Which Is Being Reclaimed by the Reclamation Service Under the "Yuma Project.* Yuma: Press Morning Sun, 1915.

4. L. C. Shattuck to Byron Pattison, August 25, 1914.

5. The 320 acre tract in section 15 was purchased for $75 an acre; the 160 acres in section 20 was bought for $52.50 per acre.

6. Yuma *Morning Sun*, October 13, 1915, p. 4.

7. W. F. Timmons to L. C. Shattuck, April 20, 1916; L. C. Shattuck to W. F. Timmons, April 22, 1916.

8. L. C. Shattuck to A. E. Deyo, July 24, 1916.

9. A. E. Deyo to L. C. Shattuck, July 31, 1916; L. C. Shattuck to A. E. Deyo, August 7, 1916.

10. A. E. Deyo to L. C. Shattuck, August 29, 1916.

11. For discussion of development of hybrid cotton at Yuma and elsewhere in the Southwest consult: Joseph C. McGowan, *The History of Extra-Long Staple Cottons*. The University of Arizona, 1960. Master's Thesis.

12. Articles of Incorporation of the Winterhaven Commercial Company, September 29, 1917.

13. A. E. Deyo to L. C. Shattuck, December 30, 1916.

14. L. C. Shattuck to A. E. Deyo, February 10, 1917.

15. Ibid., February 6, 1917.

16. Ibid., June 6, 1917.

17. A. E. Deyo to L. C. Shattuck, February 8, 1916; E. G. Attaway to A. E. Deyo, May 29, 1917.

18. L. C. Shattuck to R. M. Moore, August 7, 1917.

19. A. E. Deyo to L. C. Shattuck, November 12, 1917.

20. Ibid., November 12, 1917.

21. Ibid., October 31, 1917.

22. L. C. Shattuck to J. P. Glass, June 20, 1918.

23. Ibid., April 7, 1918.

24. Ibid., April 7, 1918.

25. L. C. Shattuck to A. E. Ketter, April 13, 1918.

26. A. E. Ketter to L. C. Shattuck, July, 15; July 22, 1918.

27. L. C. Shattuck to J. P. Glass, July 15, 1918.

28. L. C. Shattuck to A. E. Ketter, July 30, 1918.

29. Bob Cunningham, "Sanguinetti: Moses or Midas?" *Journal of Arizona History*, Volume 23, Number 2, Summer 1982, pp. 187–208.

30. L. C. Shattuck to James H. Maxey, August 15, 1918.

31. A. E. Ketter to L. C. Shattuck, October 19, 1918.

32. Ibid., January 23, 1919.

33. L. C. Shattuck to A. E. Ketter, January 28, 1919.

34. Ibid., January 29, 1919.

35. I. E. Burgess to L. C. Shattuck, March 1, 1919.

36. A. E. Ketter to L. C. Shattuck, March 24, 1919.

37. W. D. Drake to L. C. Shattuck, April 23, 1919.

38. A. E. Ketter to L. C. Shattuck, May 12, 1919.

39. L. C. Shattuck to A. E. Ketter, April 12, 1919; S. Clemens to L. C. Shattuck, April 7, 1919.

40. I. E. Burgess to L. C. Shattuck, n.d.; L. C. Shattuck to I. E. Burgess, July 16, 1919.

41. P.M. Buckwalter to L. C. Shattuck, December 5, 1919.

42. L. C. Shattuck to W. A. Markham, September 6, 1919.

43. These sales were approved by the board of directors of the Winterhaven Commercial Company in a special meeting called on December 18, 1919, to devise ways and means of meeting the company's indebtedness to the Miners and Merchants Bank. See Minutes of Winterhaven Commercial Company.

44. Minutes of Special Stockholders Meeting, December 4, 1922.

45. L. C. Shattuck to L. L. Odle, November 14, 1919.

Chapter Eleven

1. Bisbee *Daily Review*, January 14, 1921, p. 5.

2. Ibid., January 8, 1921, p. 2.

3. Ibid., January 23, 1921, p. 6.

4. L. C. Shattuck to Rhoda Leavell, April 17, 1922.

5. L. C. Shattuck to Jack Graham, March 29, 1921.

6. L. C. Shattuck to Max Marks, September 3, 1923.

7. L. C. Shattuck to Warner Shattuck, April 8, 1924.

8. L. C. Shattuck to Ely Lilly & Company, May 14, 1923; Shattuck to Dr. Chenoweth, June 22, 1923.

9. Bisbee *Daily Review*, October 24, 1924, p. 2; L. C. Shattuck to Rosa Bonnell, February 26, 1925.

10. Morris Goldwater to L. C. Shattuck, October 28, 1924.

11. L. C. Shattuck to George A. Walker, October 27, 1924.

12. L. C. Shattuck to O. M. King, January 16, 1925.

13. L. C. Shattuck to Spencer Shattuck, February 28, 1925.

14. Ibid.

15. Spencer Shattuck to L. C. Shattuck, December 20, 1924.

16. Spencer Shattuck to L. C. Shattuck, December 24, 1924. In this letter Spencer suggested to his father, "Say Dad why in hell don't you offer him a job to work the Estancia for a while. Open up and make him this offer and pay him a salary of $200. . . . The money don't mean so much to you, dad, and it will help him. . . . This is the only way to get him in a place where we will know where he is, otherwise he will wander all over as he used to do."

17. Mark Shattuck to L. C. Shattuck, July 26, 1925; L. C. Shattuck to Mark Shattuck, August 27, 1925.

18. Isabel Shattuck to L. C. Shattuck, August 27, 1925; L. C. Shattuck to Isabel Shattuck, September 1, 1925.

19. L. C. Shattuck to Spencer Shattuck, October 3, 1925.

20. Mark Shattuck to L. C. Shattuck, October 5, 1925.

21. Ibid., April 27, 1925.

22. L. C. Shattuck to Jojn Treu, May 26, 1927.

23. Bisbee *Daily Review*, November 26, 1926, p. 1, sec. 2.

24. Telegram, L. C. Shattuck to Mrs. Earl Newcomer, October 13, 1928; see also *Daily Review*, October 13, 1928, p. 8.

25. L. C. Shattuck to E. Martinelli, May 25, 1927.

26. Isabel Shattuck to L. C. Shattuck, n.d. [June 1929].

27. Mark Shattuck to L. C. Shattuck, August 16, 1929.

28. Development of the Denn was financed by the company's directors contributing $400,000, as follows: Thomas Bardon and L. C. Shattuck each contributed $150,000, and H. L. Mundy and Byron Pattison contributed $50,000 each. See Norman E. LaMond to L. C. Shattuck, December 17, 1928.

29. L. C. Shattuck to Walter J. McGurdy, August 7, 1929.

30. F. O. Mackey to L. C. Shattuck, September 18, 1922; Shattuck to Mackey, September 20, 1922; Shattuck to Mackey, April 22, 1926; Mackey to Shattuck, April 20, 1926.

31. E. A. Tovrea to L. C. Shattuck, January 3, 1923; Shattuck to Tovrea, January 23, 1923.

32. W. H. Shattuck to L. C. Shattuck, November 21, 1921; L. C. Shattuck to W. H. Shattuck, December 15, 1921.

33. William Riggs to L. C. Shattuck, December 22, 1921; L. C. Shattuck to William Riggs, January 3, 1922.

34. B. A. Packard to L. C. Shattuck, March 23, 1926; Shattuck to Packard, March 24, 1926.

35. William E. Hough to L. C. Shattuck, May 31, 1927; L. C. Shattuck to William E. Hough, June 3, 1927.

36. L. C. Shattuck to Mark Shattuck, May 26, 1925.

37. R. C. Jamison to L. C. Shattuck, November 14, 1927; Shattuck to D. C. Peacock, February 18, 1928.

38. D. C. Peacock to L. C. Shattuck, October 13, 1928.

39. L. C. Shattuck to Jamison & Peacock, November 27, 1928.

40. Mark Shattuck to L. C. Shattuck, October 21, 1929.

41. L. C. Shattuck to Thomas McGrath, November 1, 1929.

42. L. C. Shattuck to Wiley Fitzgerald, November 1, 1929.

43. L. C. Shattuck to P. M. Buckwalter, December 27, 1929.

Chapter Twelve

1. L. C. Shattuck to Daisy Shryrock, October 17, 1931.

2. L. C. Shattuck to Leslie Brown Fox, November 16, 1931.

3. L. C. Shattuck to Wiley Fitzgerald, October 31, 1931.

4. Thomas Bardon to L. C. Shattuck, February 29, 1932; Bisbee *Daily Review*, March 17, 1932, p. 1.

5. Bisbee *Daily Review*, January 17, 1932, p. 5.

6. Ibid., January 25, 1934, p. 3.

7. L. C. to Wiley Fitzgerald, March 30, 1931; Bisbee *Daily Review*, February 14, 1932, section 2, p. 1.

8. Douglas *Dispatch*, January 24, 1931, p. 6.

9. L. C. Shattuck to Wiley Fitzgerald, February 9, 1932.

10. Bisbee *Daily Review*, October. 21, 1934, section 2, p. 1.

11. L. C. Shattuck to Ida Lefler, October 23, 1934.

12. Hoval Smith showed up in Bisbee about 1900. His energy in promoting Calumet and Arizona Mining Company ventures, namely the Junction Mine and the Warren Realty and Development Company, earned him the title of "father" of the lower end of the Warren District. Prominent in Republican politics, Smith was regarded as one of the best political organizers in Arizona. Working out of a small office in Bakersville, Smith plotted the 1908 defeat of Marcus A. Smith. Hoval Smith biographical file, clipbooks, Arizona Historical Society, Tucson.

13. Draft of undated statement by L. C. Shattuck, in Fathauer Collection.

14. George W. P. Hunt to L. C. Shattuck, May 20, 1931; *EMJ*, Vol. 131, No. 10, May 25, 1931, p. 480.

15. *EMJ*, Vol. 131, No. 11, June 8, 1931, p. 530.

16. George W. P. Hunt to L. C. Shattuck, May 20, 1931.

17. Thomas Bardon to L. C. Shattuck, June 10, 1932.

18. Robert P. Browder and Thomas G. Smith, *Independent: A Biography of Lewis W. Douglas*. (New York: Alfred A. Knopf: 1986), pp. 63–65.

19. L. C. Shattuck to Cleve W. Van Dyke, March 23, 1932.

20. Bisbee *Brewery Gulch Gazette*, August 26, 1932.

21. Quoted in Browder and Smith, p. 64.

22. Thomas Bardon to L. C. Shattuck, June 10, 1932.

23. *EMJ*, Vol. 134, No. 7, August 1933, p. 313.

24. L. C. Shattuck to H. O. King, November 16, 1933.

25. Ibid.

26. *EMJ*, Vol. 135, No. 4, April 1934, p. 178–79.

27. L. C. Shattuck to C. E. White, March 18, 1931; L. C. Shattuck to Cornelius O'Connell, December 10, 1930.

28. L. C. Shattuck to B. M. Pattison, April 23, 1932.

29. L. C. Shattuck to E. F. Sanguinetti, July 18, 1930.

30. Emil C. Eger to L. C. Shattuck, March 3, 1930.

31. B. A. Packard to L. C. Shattuck, December 22, 1930; Ibid., February 2, 1931.

32. L. C. Shattuck to B. A. Packard, March 18, 1931.

33. E. F. Sanguinetti to L. C. Shattuck, July 4, 1930.

34. L. C. Shattuck to E. F. Sanguinetti, July 18, 1930.

35. Bisbee *Daily Review*, June 24, 1932, p. 3.

36. L. C. Shattuck to B. B. Moeur, March 3, 1933.

37. Bisbee *Daily Review*, March 3, 1933, p. 1.

38. Walter Reed Bimson, of the Valley National Bank, claimed his bank saved the Riggs' institution. Jim Gentry, Shattuck's lawyer, contradicts that assertion. Gentry often related how he and an armed guard transported a satchel of money from Bisbee to Willcox. For Bimson's story see Larry Schweikart, *A History of Banking in Arizona.* (Tucson: University of Arizona Press, 1982), p. 95.

39. Larry Schweikart, pp. 110–11.

40. L. C. Shattuck to Dan Angius and John Riggs, June 7, 1933.

41. Resources of the Bank of Bisbee declined from over $3 million in December of 1932 to $1,859,201 in December 1933. In contrast the Miners and Merchants Bank reported $3,825,403 total assets at end of 1933. Bisbee *Daily Review*, January 14, 1934, p. 3; March 5, 1934, p. 3.

42. Ibid., February 1, 1934, p. 1.

43. Ibid., November 22, 1933, p. 3; November 23, 1933, p. 8.

44. Ibid., April 11, 1933, p. 3.

45. Walter Fathauer to L. C. Shattuck, January 11, 1932.

46. L. C. Shattuck to Dorothy Shattuck Stensland, April 22, 1932.

47. Mark Shattuck to L. C. Shattuck, January 7, 1935. Mark attributed the advance of Denn stock to "small pooling" of some New York stockholders.

48. L. C. Shattuck to Rufus E. Patterson, June 18, 1935.

49. L. C. Shattuck to Leonard Koch, July, 30, 1934.

50. L. C. Shattuck to Wiley Fitzgerald, Sept. 1, 1934.

51. *EMJ*, Vol. 136, No. 12, December 1935, p. 625.

52. L. C. Shattuck to D. C. Peacock, April 22, 1935.

53. L. C. Shattuck to Abraham D. Ortiz, November 7, 1934.

54. L. C. Shattuck to Neil Erickson, Jan. 23, 1933.

55. L. C. Shattuck to Wiley Fitzgerald, December 8, 1937.

56. Ibid., Wiley Fitzgerald, Jan. 12, 1937.

57. L. C. Shattuck to Isabel Fathauer, December 19, 1936.

58. L. C. Shattuck to Dorothy Stensland, February 3, 1938.

59. L. C. Shattuck to Wiley Fitzgerald, April 6, 1937.

60. L. C. Shattuck to H. C. Johnson, September 16, 1937.

61. *EMJ*, 142, No. 4, April 1941, p. 84.

62. Ibid., Vol. 146, No. 4, April 1945, p. 133.

63. *Southwestern Pay Dirt*, September 1987, p. 4A; and December 1990, p. 7A

64. Bisbee *Daily Review*, July 30, 1952, p. 3.

INDEX